PLATE I.

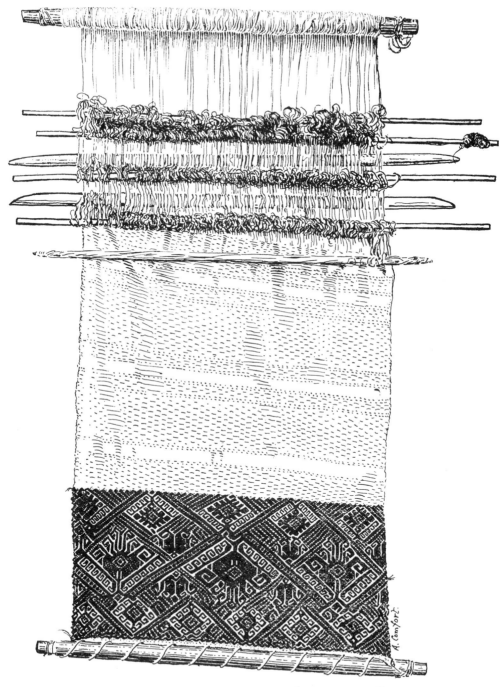

MAZATEC LOOM.   BRITISH MUSEUM (J. COOPER CLARK).

H. LING ROTH :   STUDIES IN PRIMITIVE LOOMS,

Studies in Primitive Looms

by H. Ling Roth

Reprinted from the original
Bankfield Museum Notes,
Halifax, England.

RUTH BEAN
CARLTON, BEDFORD
1977

Ruth Bean, Victoria Farmhouse, Carlton, Bedford, England
1977

ISBN   0 903585 03 0

Reprinted by permission of the original Publishers.

A VOTE OF THANKS.

To all those who have in any way assisted in making me comfortable in the course of my investigations, viz. :—Sir C. Hercules Read, for permission to make use of the Collections in his Department in the British Museum, and for calling my attention to the Korean Drawings by Becher ; likewise to his assistants, Mr. Reginald A. Smith and Capt. T. A. Joyce, and to the following Keepers of Museums for similar kindness : Dr. H. S. Harrison, Horniman Museum ; Baron A. von Huegel, Cambridge Museum of Archæology and Ethnology ; Mr. F. Leney, Norwich Castle Museum ; Mr. E. E. Lowe, Leicester ; Mr. Toms, Brighton ; Dr. S. E. Chandler, Imperial Institute Museum ; Dr. Club and his assistant, Mr. Entwhistle, Liverpool ; Dr. W. H. Tattersall, Manchester University Museum, as well as Mr. T. A. Coward and Miss W. M. Crompton ; Prof. Wm. Myers, Manchester Municipal School of Technology ; Mr. Ben H. Mullen, Salford ; The Director, Royal Scottish Museum, Edinboro' ; Mr. Campbell, Ethnographical Department, Glasgow Museum ; Dr. W. H. Miller, Dundee, and the Director, Victoria and Albert Museum.

I also wish to thank Dr. Haddon for assistance in the loan of Books ; Messrs. Bulleid and Gray for permission to reproduce Fig. 183 from their work on the Glastonbury Lake Village ; Mr. I. H. N. Evans for information regarding Ilanum Looms ; Dr. A. D. Imms ; Mr. J. A. Hunter, author of " Wool " ; Miss M. L. Kissel for troubling to get me information regarding the Hopi pressers-in in the New York Natural History Museum ; Miss B. Freire-Marreco for generously placing her knowledge of Hopi Weaving at my disposal, and Miss Laura E. Start for various valued suggestions.

H. LING ROTH.

HALIFAX, YORKS.,
15th October, 1918.

PREFACE TO THIRD EDITION

" Studies in Primitive Looms " was written by H. Ling Roth between 1916 and 1918, when he was Keeper of the Bankfield Museum, Halifax.

The continued demand for this publication necessitated a reprinted edition in 1934 and this further edition in 1950. The present edition comprises the original text with the addition of a more comprehensive index.

Grateful acknowledgements are tendered to Messrs. Kingsley Roth and Alfred B. Roth, sons of the late H. Ling Roth, for permission to undertake the reprinting, and to the Council of the Royal Anthropological Institute for the gift of the blocks.

R. PATTERSON.

BANKFIELD MUSEUM, HALIFAX.
January, 1950.

CONTENTS

STUDIES IN PRIMITIVE LOOMS

By H. Ling Roth.

I.—Introduction and Definition of Terms.

A GREAT deal has been written about primitive weaving tools, and if I add to the quantum it is partly because I venture to think I have something new to say, and partly because I wish to bring to the notice of traveller and students at home and abroad the necessity for gathering further, and above all correct, information on the subject before it is too late. There is nothing so annoying as the crude descriptions we are supplied with, when a little care could and should bring us invaluable knowledge ; and as to the illustrations, the authors seem to hide just that which is the most important for us to see. Here in England, in the greatest textile-producing country in the world, we still evince little interest in the subject.

Weaving is generally considered to be the outcome of basketry and mat-making, and in most cases probably it is so. It consists of the interlacing at right angles by one series of filaments or threads, known as the *weft* (or *woof*) of another series known as the *warp*, both being in the same plane.

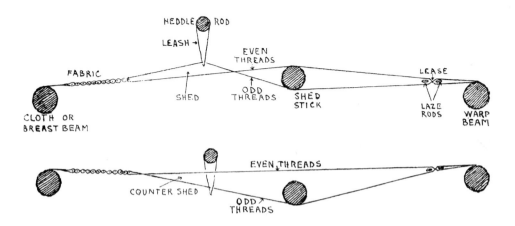

DIAGRAM TO ILLUSTRATE THE PRINCIPLES OF WEAVING

I

The warp threads are stretched side by side from a *cloth*, or *breast-beam*, to another beam known as the *warp-beam*, often spoken of as *the beam*, and the weaving is encompassed as follows (see Fig. 1) : the *odd threads* (1, 3, 5, 7, 9, etc.) are raised by means of the fingers, leaving the *even threads* (2, 4, 6, 8, 10, etc.) in position. By raising the odd threads only, a space or opening is formed between the two sets of threads, which is called the *shed*. Through this shed the weft thread is passed, or as it is termed, a *pick*, or *shot*, is made. This weft thread (or pick) is straightened and pressed home into position at right angles to the warp by means of a *sword*, or *beater-in*. The odd threads are then dropped back into position, and the even threads are now raised instead, whereby a new or *countershed* is produced and the pick made as before. It will be understood that, as a consequence of the lifting and dropping of the odd and even threads, these two sets of threads cross each other, but remain in their respective vertical planes. This crossing keeps the pick in position.

To make the work easier and more expeditious a rod, the *heddle-* or *heald-rod*, is placed across the warp ; to this rod the odd warp threads are lightly attached by a series of *loops* or *leashes*, so that when the rod is raised all the odd threads are raised *together* instead of singly by the fingers, and through the shed so formed the pick is made. When the rod is dropped, the odd threads fall back into position between the even threads. But as the even threads are now not raised, the odd threads must be made to fall below the even threads to make the next or counter-shed. The odd threads are therefore pulled down at first by the fingers, and in the countershed so made a thick rod or *shed-stick* is inserted. This shed-stick remains in this position until the whole warp is used up, or, in other words, the piece of cloth is woven, and its action may be described as follows : when the heddle is raised, the pick made, and the heddle dropped again, the shed-stick, owing to its thickness, forces the odd threads below the even threads, and so the countershed is obtained. Later on a flat stick has come into use, which is kept flat to the warp when the heddle is raised, but set on edge when the heddle is dropped, whereby the shed is enlarged and the pick facilitated. Later still, double heddles and counter heddles, with their *harness* and *treadles*, were introduced, but these can be dealt with as they arise.

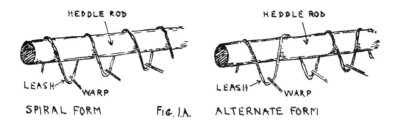

SPIRAL FORM   FIG. 1A.   ALTERNATE FORM

The heddle leashes are either single or continuous. If single (that is, if every leash is made of a separate piece of filament, spun or non-spun), the leashes are often bunched together, as in the African raphia looms, or every leash is tied up separately, as in more advanced looms. If continuous (that is, one long filament serves for making all the leaves required), then the leashes are either *spiral* or *alternate*, as shown in the illustration Fig. IA. " Spiral " means that the filament is wound loosely round the heddle-stick, and " alternate " means that the filament laps over the sides of the heddle-rod alternately.

The difficulty experienced in keeping the warp threads from getting entangled one with another, especially when these threads are long and the cloth to be woven is broad, is overcome by crossing them with one or two pairs of rods. The odd threads (more or less close to the warp beam) are raised, and one rod passed through the shed ; then the even threads are raised and the other rod passed through. This arrangement causes friction, and the warp threads are unable to move laterally, and hence retain their position and do not tangle. This crossing of the warps is called a *lease*, and hence the rods are called the *lease-rods*, corrupted into *laze-rods*. Laze-rods are, in so far as my studies go, found at the present day on nearly all primitive looms, although in the quite early stages the warps are more or less bunched at the lease, and do not require any laze-rods.

Another method of keeping the warp threads in position is the *warp-spacer*, known also as a *raddle*. It appears to have been in use in Egypt.[1] It is provided with pegs or teeth, between which the warp threads are passed in various definite quantities. The space *between* the teeth or pegs is called a *dent* by weavers, although the loom-makers call the tooth or peg the dent ; as we are dealing with weaving and not with machine-making, it will be as well to adhere to the users' definition.

In course of time the beater-in was supplanted by a comb-like article which developed into the *reed* and later still into the *sley*, a tool which drives home the weft as well as keeps the warp in position. I say " supplanted " advisedly, as so far I cannot trace any evolution in the matter, and, judging by specimens of reeds from the Philippines and Borneo, the reed was originally a form of warp-spacer, and ultimately became a beater-in as well. But in any case the reed appears to have made its appearance very late.

There is a third method which consists in fixing the warp threads separately on the beams by means of a heading or tailing thread, but this is only effective on short looms.

Now as to the meaning of the word *loom*. According to the *New Oxford Dictionary* it is of obscure origin, and meant in the first instance " an implement or tool of any kind," now applied to " a machine in which yarn or thread is woven into fabric by the crossing of threads called respectively the warp and weft." Some writers only apply the term loom to the frame when it refers to weaving in which the shed is no longer obtained by means of the fingers (or a pointed stick or a spool point), but by mechanical means, viz., the heddle. But I think that as long as a fabric, *i.e.*, anything woven in the accepted signification of the term, is obtained

---

[1] See *Ancient Egyptian and Greek Looms*, by H. Ling Roth, Halifax, 1913, p. 20.

3

the frame on which it is obtained had better be called a loom, and in that sense I use it in these studies.

2.—The Evolution of the Spool and Shuttle.

There seems some confusion as to what is a spool or bobbin, and as to what is a shuttle, nor is it at first sight quite easy to draw a hard and fast line. I should describe a spool or bobbin as a quill or small cylindrical shaft on which the weft is wound for the purpose of weaving, and a shuttle as an instrument for the same purpose, consisting ultimately of a more or less boat-shaped *case containing a spool*. In the accompanying diagram (Fig. 2) I have made an attempt to portray the evolution of both from a single short filament. The lines of evolution seem to be three :

(*A*) One in which the filament is wound round the spool more or less lengthwise, *i.e.*, parallel to the axis of the spool ;

(*B*) the other in which the filament is wound more or less at right angles to the axis of the spool ; and

(*C*) the third, in which the attachment can be likened to the threading of a needle, as in the Iceland specimen, or where instead of a needle eye there is a slot, as in the African beater-in and spool combined. As regards this African tool, the slot points away from the body or blunt end of the tool, hence it would appear that it is pushed through the shed, and when it emerges at the other side the filament is put into the slot and the spool withdrawn the way it entered, leaving the filament in its place. There are some more advanced forms of the African beater-in and spool combined, which have the slot pointing both ways.

Evolution along the line *A* : from winding the weft lengthwise and covering the spool almost to its very ends, which are rounded or cut off straight at first, we find the ends become grooved as in the Slave Indian specimen *Ab*, and gradually as the grooves deepen the now double ends appear to lengthen, *Ab1*, and take the shape of horns as with the Ainu spool *Ab2* ; then these double ends incline towards each other like those of our fishermen's mesh-pin or needle, and finally recurve backwards as in the second Ainu example *Ab3*.

At the same time, as a branch development, the rounded sides of the spool become flattened and in turn become grooved so that there is a longitudinal groove as well as an end one, that is to say, the groove is continuous ; this is seen in the Santa Cruz specimen *Ad1*, and still more so in the Iban specimen *Ad2*, which is practically the same, only on a larger scale, as our well-known English ladies' tatting shuttle, so called.

On the line of evolution *B* with its transverse winding at right angles to the spool, the resultant bulginess attained by this method of winding may have necessitated a cover or case to facilitate the making of the pick. As a first form of such a case we have in the Malay specimen *Bb* a piece of cane cut off at a node

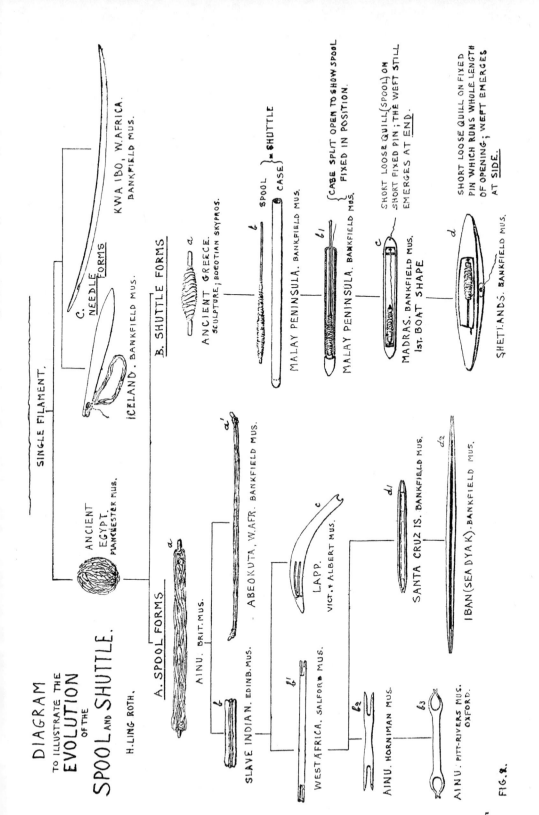

DIAGRAM
TO ILLUSTRATE THE
EVOLUTION OF THE
SPOOL AND SHUTTLE.
H. LING ROTH.

SINGLE FILAMENT.

C. NEEDLE FORMS

KWA IBO, W. AFRICA. BANKFIELD MUS.

ICELAND. BANKFIELD MUS.

ANCIENT EGYPT. MANCHESTER MUS.

B. SHUTTLE FORMS

a. ANCIENT GREECE. SCULPTURE; BOEOTIAN SKYPHOS.

b. SPOOL ⎫ = SHUTTLE
   CASE ⎭
MALAY PENINSULA. BANKFIELD MUS.

b¹. {CASE SPLIT OPEN TO SHOW SPOOL FIXED IN POSITION.
MALAY PENINSULA. BANKFIELD MUS.

c. SHORT LOOSE QUILL (SPOOL) ON SHORT FIXED PIN; THE WEFT STILL EMERGES AT END.
MADRAS. BANKFIELD MUS. 1st. BOAT SHAPE

d. SHORT LOOSE QUILL ON FIXED PIN WHICH RUNS WHOLE LENGTH OF OPENING; WEFT EMERGES AT SIDE.
SHETLANDS. BANKFIELD MUS.

A. SPOOL FORMS

a. AINU. BRIT. MUS.

a¹. ABEOKUTA, W. AFR. BANKFIELD MUS.

b. SLAVE INDIAN. EDINB. MUS.

c. LAPP. VICT. & ALBERT MUS.

d. IBAN (SEA DYAK). BANKFIELD MUS.

d¹. SANTA CRUZ IS. BANKFIELD MUS.

d². SANTA CRUZ IS. BANKFIELD MUS.

b¹. WEST AFRICA. SALFORD MUS.

b². AINU. HORNIMAN MUS.

b³. AINU. PITT-RIVERS MUS. OXFORD.

FIG. 2.

5

which has been slightly rounded.   In this the spool is placed, not fixed, and the weft unravels from the open end.   A development of this consists in making the blunt end pointed, or making a point of gum or of some other resinous substance, as in another Malay specimen not shown.   Later on this point is replaced by a wooden stopper with an internal socket, into which the spool is rammed, and so gets fixed as in a third Malay specimen $Bb1$.   Then to ease the unravelling in the tube the spool is shortened, for the case is now held by the hand instead of the spool being held, and further to assist supervision the upper longitudinal half of the cane case appears to have been cut away, the end plugged and perforated, to guide the outgoing weft, and we get the first shuttle as known to us in the boat form with the weft still running out endwise as in the Madras specimen Bc. There is, however, an objection to the end outlet, inasmuch as a shuttle so provided has to be turned round at the end of every pick, hence the outlet for the weft was made at the side instead, while the spool was made to fit the full length of the case opening in the shuttle Bd, and this is the general form the shuttle practically retains at the present day.

The longitudinal method of winding the thread on the bobbin, i.e., method A, is evidently due to the fact that owing to its slender shape the spool passes easily through the shed.   On the other hand, the thickness resulting with method B makes the pick more difficult, a difficulty which was overcome by the use of a case, the adoption of which was facilitated by the shortening of the spool.

In order to test this I had a wooden spool made 22 inches (or 56 cm.) long and $\frac{9}{16}$ inch (or 14 mm.) in diameter, and had it carefully wound round with 200 yards of thread according to method A ; this covered the spool to a length of 21 inches (or 53 cm.) and increased its diameter to $\frac{3}{4}$ inch (or 19 mm.).   Then another spool of wood was obtained $8\frac{1}{2}$ inches (or 21 cm.) long, and also $\frac{9}{16}$ inch (or 14 mm.) in diameter ;   200 yards of the same thread was wound round this according to method B, covering the spool to a length of $3\frac{1}{2}$ inches (or 8·9 cm.), and increasing its diameter to $1\frac{3}{8}$ inches (or 35 mm.).   It is therefore evident that by the method A the same quantity of thread can be carried through a small shed with greater ease than can be carried through by method B.   In connection with this I find also, generally speaking, the method A, the longitudinal method, is in use only with the more primitive forms of loom, i.e., those in which the countershed is still made by a shed stick, and where consequently the shed is not so clear, nor opened so widely, as in looms provided with counter heddles and treadles.   In other words, the improvement in the loom permitted improvement in the spool, which led to the evolution of the shuttle.

According to the explanation given in the *New Oxford Dictionary*, the origin of the word bobbin is unknown, but we are informed that Cotgreve, writing in 1611, calls it " a quil for a spinning wheele."   The word spool is merely another term for bobbin, is of Teutonic origin, and is also applied " to the mesh-pin used in net-weaving."   As regards the word shuttle, the same dictionary tells us that primarily it meant a " dart, missile, arrow " an explanation which appears to me to designate correctly the quality which distinguishes it from the spool, for I have found it easier to shoot through a shed with a shuttle than with a spool. It was no doubt this shooting capacity which led Kaye to invent the flying shuttle in 1733.

To come back to the Dictionary, we find it says : "The normal form of the shuttle resembles that of a boat, whence its name in various languages (L. *navicula*, F. *navette*, G. *Webeschiff*)." As regards the French and German interpretations the Dictionary is most probably correct, but not so as regards the Latin interpretation. Asking Professor T. F. Tout, Manchester University, for assistance in the explanation of the extended meaning of the word *navicula*, he kindly replied, saying, " *Radius* and *pecten* are the ordinary classical words for shuttle," and quoted the well-known lines in Virgil : *Arguto tenues percurrens pectine telas* (as she runs over her delicate web with the nimble spool), *Æn.*, VII, 14. He continued, " The ordinary dictionaries, *e.g.*, Lewis and Short, do not give *navicula* in the sense of shuttle at all. Ducange, s.v. *navicula*, quotes Ugotio (a twelfth-century writer, I *think*), *Radius instrumentum texendi, scilicet pecten vel navicula* (Bobbin, an instrument for weaving, that is a quill or shuttle), a good passage for your purpose. The French *navette* is also used in the thirteenth century in its modern sense of shuttle. This is as far back as my reference books give the word. If one had time, no doubt, earlier instances could be found, but *navicula* is certainly post-classical for shuttle, though probably earlier than the twelfth century. This helps your point that early shuttles were not like little boats."

Blümner, in his great work,[1] after describing κερκίς (spool), as used by the Greeks, continues : " But, apart from this, Homeric times seem to have known the real shuttle, in which the weft is wound round a spool inside and unravels through an opening in the shuttle when it is thrown. In the above-mentioned passage in Homer—viz., *Iliad*, XXII, 760—it is stated of the female weaver : πηνίον ἐξέλκουσα παρὲκ μίτον (drawing the spool across the web) ; here πηνίον takes the place of κερκίς. This πηνίον we also meet with elsewhere, yet it cannot be considered identical with κερκίς, but is explained by acknowledging that the spool within the shuttle is referred to." But why ? I ask. The passage is clear enough without such an inference. Blümner then quotes various forms of πηνίον and its application, but there is not in any one of his quotations any description which can in the remotest way be applied to a shuttle. He also quotes in support of his contention the well-known post-classical fourth century A.D. lexicographer Hesychius, who most probably had seen a shuttle at the time of his writing ; but while this authority does distinguish between the spool and the shuttle, this does not prove the existence of the shuttle some 1100 or 1200 years before his time. Marquardt[2] says πηνίον is the weft (the *Eintragfaden*), and this fits the passage well enough, " drawing the weft across the web " ; spool and weft are equally correct. It seems as

[1] *Techn u. Term. der Gewerbe u. Kuenste bei Griechern u. Roemern*, 1st vol., 2nd ed., Leipzig, 1912, pp. 152-3.
[2] *Das Privatleben der Roemer*, Leipzig, 1879, p. 504, note 7.

though Blümner, having given his opinion that in Homer's time the shuttle was known, is attempting to give a new reading to the word πηνίον in order to sustain his contention. He is, however, misled by an illustration (the bottom central one of Fig. 160 on p. 157 of the *British Museum Guide to the Exhibition illustrating Greek and Roman Life*) which he takes to be a shuttle (*Webeschiffchen*), but which is in reality the grooved spool of type *Ad*1 of my diagram, and most decidedly *not* a shuttle.

The noise made by the shuttle and referred to in the classics is also brought in as a proof of the existence of the tool at the period named. A purposely obscure passage in Aristophanes' *Frogs*, 1316, quoted by Marquardt, p. 509, note, reads :—

ἱστότονα πηνίσματα

κερκίδος ἀοιδοῦ μελέτας,

"the weft stretched on the web beam—the care of the tuneful shuttle." What is a tuneful shuttle ? And in the quoted passage from the *Æneid* we are treated to the "singing" of the shuttle. In this case we have an alternative for *arguto* (singing), by translating it "deft" or "nimble," and it would appear that if "nimbleness" be accepted, it must be on account of poetical licence—and we are dealing with poets—in which the deftness of the weaver is transferred to her implement. On the other hand, there appears no other meaning for ἀοιδός than "tuneful," and with regard to the low state of musical culture among the Greeks, what may have been tuneful to them is most probably not tuneful to us. The fact is, a noise was made during weaving and recorded, the recorder not being very precise as to whether the noise emanated from the tool which set the work in motion or from the loom. I have ascertained by experiment on various more or less primitive looms in Bankfield Museum, that some shuttles make no noise, while others do, and that, generally speaking, whether spools or shuttles are used, the noise the observer notes comes from the loom itself and not from the shuttle.

In their earlier periods the Greeks had vertical looms, with warp weights, which possibly in later times were replaced by a lower or breast beam. As explained above, the shuttle evolved with the improvements of the loom. It evolved as other things evolve, as the opportunity or necessity for it arose. There was little opportunity, if any, and practically no necessity, for the shuttle on the warp-weighted loom of so primitive a construction as that possessed by the earlier Greeks. Hence, taking all the points into consideration, it appears to be an anachronism to infer that the shuttle existed in Homeric, or perhaps even in later Greek, times. What was used was still a spool.[1]

---

[1] Since the above was written I find that Otto Schrader (*Lingu.-hist. Forschungen zur Handelsgeschichte u. Warenkunde*, Jena, 1886, p. 182), as quoted by Franz Stuhlmann (*Ein Kulturgeschichtlicher Ausflug in den Aures* . . . Hamburg, 1912, p. 195), states : "Our shuttle was unknown to the Ancients." I have not been able to see a copy of Schrader's work,

3.—THE AINU LOOM.

The Ainu loom is a primitive affair, with characteristics well worth studying. It has not been described before to any extent except in a very crude and unsatisfactory way by Hugo Ephraim,[1] and hence I have chosen it as a fit subject for discussion.

FIG. 3.—AINU WOMAN WEAVING.
(After Romyn Hitchcock.)

FIG. 4.—AINU WOMAN WEAVING.
(As reproduced by Ephraim, after Romyn Hitchcock. Note the distorted heddle and spool, and the gratuitous and incorrect addition of the feet.)

Both reduced by one-fourth lineal.

We are told by the Rev. John Batchelor that " the chief article of dress worn by the Ainu is a long garment which they call *attush*. This word really means elm fibre or elm thread, and, as the words indicate, the dresses are made from the inner bark of elm trees. Such garments are very brittle when dry, but when wet they are exceedingly strong."[2] According to MacRitchie, in the legend descriptive of the illustration of Ainu peeling the bark off the tree, it is " *Microptelea parvifolia*, in Aino *ohiyo*," but Hitchcock says the people use the bark of the elm (*Ulmus montana*), called by them *ohiyo*, and sometimes *U. campestris*.[3] Batchelor continues : " Elm bark is peeled off the trees in early spring or autumn, just when the sap commences to flow upwards or when it has finished doing so.[4] When sufficient bark has been taken, it is carried home and put into warm, stagnant water to soak. It remains here for about ten days till it has become soft ; then, when it has become sufficiently soaked, it is taken out of the water, the layers of bark separated, dried in the sun, and the fibres divided into threads and wound up into

---

[1] " Ueber die Entwickelung der Webetechnik . . ." (*Mitt. a. d. städisch. Mus. f. Voelkerk. zu Leipzig*, 1905).

[2] *The Ainu and their Folklore*, Lond., 1901, 2nd ed., p. 144.

[3] " The Ainos of Yezo," *Rep. U.S. Nat. Mus. for* 1890, p. 463.

[4] " The men bring in the bark, in strips 5 feet long, having removed the outer coating" (Bird).

balls for future use. Sewing-thread is sometimes made in the same way, only it is chewed till it becomes round and solid. Sometimes, however, thread is made by chewing the green fibre as soon as taken from the trees. When all the threads have been prepared, the women sit down and proceed with their weaving."

Going a little more into detail, Miss Bird says : " This inner bark is easily separated into several thin layers, which are split into very narrow strips by the older women, very neatly knotted, and wound into balls weighing about a pound

Fig. 5

Fig 6

AINU WEAVING. YEZO-MANGA PICTURES —MACRITCHIE

Fig 7

AINU STRIPPING BARK OFF ELMS. ·
YEZO-MANGA PICTURES—MACRITCHIE

AINU LOOM
MATSMAE PANORAMA — MACRITCHIE

each . . . . The loom consists of a stout hook fixed in the floor, to which the threads of the far end of the web are secured, a cord fastening the near end to the waist of the worker,[1] who supplies, by dexterous rigidity, the necessary tension ; a frame like a comb resting on the ankles, through which the threads pass ; a hollow roll for keeping the upper and under threads separate, and spatula-shaped beater-in[2] of engraved wood, and a roller on which the cloth is rolled as it is made. The length of the web is 15 feet, and the width of the cloth 15 inches. It is woven with great regularity, and the knots in the thread are carefully kept on the underside. It is a very slow and fatiguing process, and a woman cannot do much more than a foot (30 cm.) a day. The weaver sits on the floor with the whole arrangement attached to her waist, and the loom, if such it can be called, on her ankles. It takes long practice before she can supply the necessary tension by spinal rigidity. As the work proceeds, she drags herself almost imperceptibly nearer the hook. In this house and other large ones two or three women bring in their webs in the morning, fix their hooks, and weave all day, while others, who have not equal advantages, put their hooks in the ground and weave in the sunshine. The web and loom can be bundled up in two minutes, and carried away quite as easily as a knitted sofa

[1] This is the semi-girdle or backband. A. S. Bickmore (*Trans. Ethnol. Soc.*, vii, N.S., 1869, p. 18) speaks of it as a board. If so it is somewhat similar to the old-fashioned Japanese backpiece.

[2] Bird uses the word " shuttle " here, but it is evidently a clerical error.

blanket.[1] Batchelor tells us the garments produced " are very rough indeed, reminding one of sackcloth, and are of a dirty brown colour. It is therefore no wonder that those Ainu who can afford it prefer to wear the softer Japanese clothing."

Simple as it looks, the Ainu loom is characteristic in all its parts except one, and this one, the semi-girdle or back-strap, appears to connect it with the looms of, generally speaking, the Pacific Region. The users of the back-strap are, or were, the Bhutiyas of N.W. India (specimen in Bankfield Museum), the Tibetans,[2] the Chinese,[3] Burmese and Assamese,[4] the Iban or Sea-Dyaks (Bankfield Museum), the Japanese,[5] Philippine Islanders,[6] the Koreans,[7] the Santa Cruz Islanders (Bankfield Museum), the Caroline Islanders, the Aztecs,[8] and Modern Mexican tribes (British Museum), and so on—a fairly wide circle of users. On the British Museum Ainu loom the back strap is of bark, on the specimens in the Horniman Museum and the Royal Scottish Museum, Edinburgh, it is of wood.

The Ainu, as observed, use non-spun bast filament in single strips both for warp and for weft. A similar non-spun filament, but in much broader strips, and on a much cruder loom, is used by the Kwakiutl Indians for mat-making.[9] The Santa Cruz Islanders use a non-spun filament for their warp, and a twisted filament or thread for the weft ; what the latter is made of I have not been able to ascertain, but the warp is said to be obtained from the stem of a black banana ![10] According to Lieut. Emmons, in the manufacture of the Chilkat blanket the inner bark of the yellow cedar (*Chamaecyparis nootkatensis*) and of the red cedar (*Thuya plicata*) is laid up in a two-stranded cord, so it is bast thread and non-spun bast filament.[11] Otherwise the great field for non-spun filament used for weaving and drawn from the Raphia palm is the vast region of that palm's habitat in Africa. But beyond using a non-spun filament, there is nothing in common between the looms of Africa and the looms of the Ainu. In a Shan head-dress in Bankfield Museum some weft is of non-spun filament-like palm leaf splittings, and native cotton warp.

There does not appear to be any warp beam in the Ainu loom, a *warp peg* (Fig. 27) being used instead, and is driven into the ground, as is evident from the specimens in the British Museum, Royal Scottish Museum, Edinburgh, and Horniman Museum. Hitchcock does not mention any warp attachment, and Bird mentions the use of a hook.

[1] *Unbeaten Tracks in Japan*, 4th ed., Lond., 1881, ii, 92.
[2] W. H. S. Landor, *Tibet and Nepal*.
[3] Falcot, *Traite*, 1852, Pl. 224.
[4] Joyce and Thomas, *Women of All Nations*.
[5] v. Bavier, *Japan's Seidenzucht*, Pl. IV, Fig. 3.
[6] Worcester, *Philippine Journ. Sci.*, I.
[7] Cavendish, *Corea and the Sacred White Mountain*.
[8] Kingsborough's *Mendoza Codex*.
[9] M. L. Kissel, *Aboriginal American Weaving*.
[10] Florence Coombe, *Islands of Enchantment*, London, 1911, p. 175.
[11] " The Chilkat Blanket," *Mem. Amer. Mus. Nat. Hist.*, iii, p. iv. Dec., 1907.

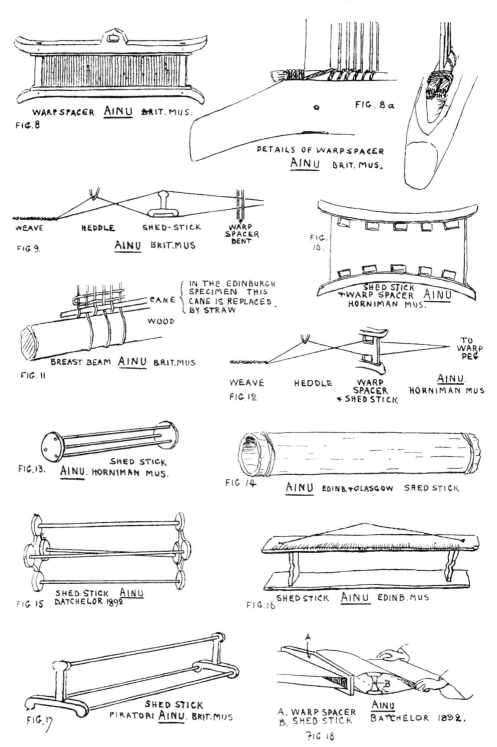

WARP SPACER AINU BRIT. MUS.
FIG. 8

FIG. 8a

DETAILS OF WARP SPACER AINU BRIT. MUS.

WEAVE   HEDDLE   SHED-STICK   WARP SPACER DENT
FIG. 9.   AINU BRIT. MUS

FIG. 10.

CANE   (IN THE EDINBURGH SPECIMEN THIS CANE IS REPLACED BY STRAW

WOOD

BREAST BEAM AINU BRIT. MUS
FIG. 11

SHED STICK + WARP SPACER AINU HORNIMAN MUS.

WEAVE   HEDDLE   WARP SPACER + SHED STICK   AINU. HORNIMAN MUS
FIG. 12

TO WARP PEG

SHED STICK
FIG. 13.   AINU. HORNIMAN MUS.

FIG. 14   AINU EDINB & GLASGOW   SHED STICK

SHED STICK AINU BATCHELOR 1892
FIG 15

SHED STICK   AINU EDINB. MUS
FIG. 16

SHED STICK PIRATORI AINU. BRIT. MUS
FIG. 17

A. WARP SPACER   AINU
B. SHED STICK   BATCHELOR 1892.
FIG 18

12

At the breast or *cloth beam* there is a heading rod (Fig. 11) as shown in the illustration ; in the Royal Scottish Museum specimen the heading rod is made up of several pieces of straw, and in the British Museum specimen it is a piece of cane.

The warp length in the British Museum specimen is 14 feet (or 4·25 m.) from beam to beam, with a width of fabric of 10½ inches (or 26·5 cm.) and about 16 picks

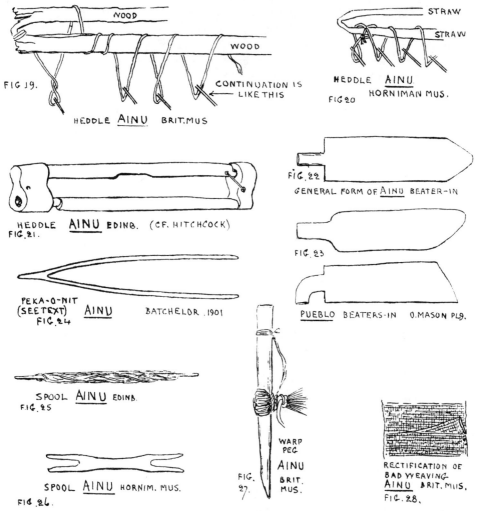

FIG 19.

HEDDLE AINU BRIT.MUS

WOOD

WOOD

CONTINUATION IS ← LIKE THIS

STRAW

STRAW

HEDDLE AINU

FIG 20 HORNIMAN MUS.

HEDDLE AINU EDINB. (CF. HITCHCOCK)
FIG. 21.

FIG. 22
GENERAL FORM OF AINU BEATER-IN

FIG. 23

PEKA-O-NIT
(SEE TEXT) AINU BATCHELOR .1901
FIG. 24

PUEBLO BEATERS-IN O.MASON PL9.

SPOOL AINU EDINB.
FIG. 25

SPOOL AINU HORNIM. MUS.
FIG 26.

FIG.
27.

WARP
PEG

AINU
BRIT.
MUS.

RECTIFICATION OF
BAD WEAVING
AINU BRIT. MUS.
FIG. 28.

to the inch (or 7 to the cm.). In the Royal Scottish Museum specimen, the warp is many feet long with a fabric width of 12¼ to 12½ inches (or 31 to 31·7 cm.). In the Horniman Museum narrow loom the length of the warp is 8 feet 10 inches (or 2·7 m.) with a fabric width of 1⅞ to 2 inches (or 5 cm.).

The warp and weft are both continuous. Bird mentions that the knots on the warp and weft are kept well out of sight ; this is, however, not always the case.

The only *pattern* I have seen in these cloths is one formed by the introduction

13

of blue, green, and white Japanese cotton warp threads in the centre of the work. Bickmore also mentions this.[1]

The Ainu horned *bobbin* (Fig. 26) is called *ahunka-mit* ; as shown in the diagram of the development of the spools it will be seen that these people follow method *A* in its various stages.

The *heddle*, for which I cannot find the native name, is of the single lifter type (Figs. 19 and 20), and like all other parts of the loom varies considerably. In the British Museum specimen it consists of a piece of stick bent double ; in the Horniman Museum specimen it is a piece of straw bent double ; in the Pitt Rivers Museum, Oxford, it is a plain cylindrical rod. In the Royal Scottish Museum, Edinburgh, one specimen (Fig. 21) is a rectangular frame made up of two shaped pieces, the leashes hanging from the lower rod, while the lower side of the upper rod is cut away for a hand grip. This form is somewhat similar to that illustrated with a woman at work, by Hitchcock (Fig. 3), but which has been distorted out of all semblance by Ephraim (Fig. 4). In the Matsmae panorama (Fig. 7), reproduced by David MacRitchie,[2] the heddle looks to be a plain rod with its leashes as in the Pitt Rivers Museum, Oxford, while in the Yezo-Manga pictures (*ibid*. Plate XVII) the rod has a bow handle (Fig. 6).

The heddle leaches are of coloured Japanese thread.

The *shed-stick*, *kamakap*, is peculiar throughout, and in some forms quite different from that used by any other peoples in so far as my enquiries go. One form (Fig. 13) consists of three cylindrical rods, which fit at their ends into circular plates as in the Horniman Museum specimen ; others, in the British Museum (Fig. 17) and Pitt Rivers Museum, Oxford, have the end plate made like an inverted $\perp$; while Hitchcock's illustration (Fig. 3) shows the end plate as an inverted $\wedge$. The Yezo-Manga (MacRitchie, Plate XVII) drawing (Fig. 6) shows three rod ends without any plate, which is evidently an oversight. Sometimes there are four rods as shown by Batchelor (Fig. 15), which fit into a fancy end plate, or they are fitted into a square plate as indicated in the Matsmae panorama (MacRitchie, Plate IX) (Fig. 7). In the Royal Scottish Museum, and in the Glasgow City Art Gallery and Museum, there are specimens of the shed-stick, made out of lengths of the stem of the rice paper plant, *Fatsie papyrifera* (Fig. 14), as kindly identified for me at the Royal Botanic Gardens, Kew—evidently similar to the one mentioned by Bird. To preserve the ends from splitting, they are bound with bast filament. Still a different form is exhibited in the Royal Scottish Museum, Edinburgh (Fig. 16). It looks like a miniature bench, and is much scored by the friction of the warp threads in making the sheds. It is shown end on in one of Batchelor's illustrations (Fig. 18), reproduced herewith. There is also a shed-stick and warp-spacer combined, and this I will refer to presently.

The Ainu evidently make mistakes like other people. In the British Museum

---

[1] *Trans. Amer. Journ. Science*, 1856.
[2] *The Ainos*, Suppl. to vol. iv. *Archiv Intern. d'Ethnographie*, 1892, Pl. IX.

specimen (Fig. 28), at a distance of about 1½ in. (or 4 cm.) from the breast beam, the weaver had got much of the weft not at right angles to the warp, owing to the faulty position of the breast beam. In order to remedy this, she has made a triangular pleat of the faulty portion, stitched it back on to the fabric, and so, getting the last few picks correctly at right angles to the warp, proceeded with her work.

We now come to another peculiar feature of the Ainu loom, viz., the *warp-spacer* or *osa*. The osa appears to exist in one of two forms, in all specimens and illustrations of Ainu looms, and in one form (Fig. 8) it resembles the well-known *reed*, which is a beater-in and warp-spacer combined. But the Ainu use this tool as a warp-spacer only, and therefore invariably place it between the warp-peg (beam) and heddle, instead of between the heddle and the fabric. When in use, two filaments (sisters) are passed through each dent. This reed-like osa, from its make and from its position, indicates want of appreciation of its double function. The other form of osa (Fig. 10) used, in the Horniman Museum specimen of an Ainu loom for making a narrow fabric is simpler in every way. It consists of a single flat piece of wood cut to shape, and is provided with one row of upper holes and one row of lower holes through which bundles of warp filaments are passed and by means of which the osa acts as a primitive warp-spacer. This perforated board-osa is incomparably simpler than the reed-like osa, and therefore most probably preceded the latter. In adopting the reed-like osa, which they probably did from outside, the Ainu seemed to have grasped the idea that it was better than their pattern, but evidently either did not grasp its use as a beater-in, or else found that there was not much benefit to be gained by adopting it with the filament they used.

It is to be observed that the Ainu do *not* make use of *laze-rods*, and are apparently among the few primitive weavers who dispense with this tool, the place of which seems to be taken by the osa. As already mentioned, the Ainu use a *shed-stick and warp-spacer combined*, and this is the board like osa (Fig. 12) in the Horniman Museum specimen. It is a combination I have not observed in any other primitive loom.

The *beater-in*, or *sword*, *attush bera* (Fig. 22) has the shape of a very broad-bladed knife ; in fact, its breadth is its distinguishing feature. I know of no other such broad beater-in on the Asiatic side of the Pacific. Otis Mason calls attention to this beater-in, whose broad batten with a handle is similar to some of those found in the Pueblo region.[1] He gives some illustrations, two of which (Fig. 23), most like the Ainu beaters-in, I reproduce. He does not say whether both edges are equally adapted for the work, as is the case with the Ainu tool. Altogether the Pueblo tools appear to be thick along one edge, and hence similar to one from ancient Peru in Bankfield Museum, and consequently not so similar to the Ainu tool as might be thought at first sight. On the other hand, the existence of the haft may be a connecting link. Batchelor illustrates a tool (Fig. 24), and says it is called *peka-o-*

[1] " A Primitive Frame for Weaving Narrow Fabrics," *Rep. U.S. Nat. Mus.*, 1899, p. 510.

nit, and is used for the purpose of changing the warp threads. Does he mean that it is a primitive heddle ? [1]

The forms of the constituents of the Ainu loom are thus seen to be in part apparently local and in part similar to those in use elsewhere. The following summary shows this more clearly, but in studying them it must be borne in mind that this investigation makes no claim to be exhaustive and that negative evidence taken by itself is always liable to be upset.

Ainu Constituents.	*Presence or Absence elsewhere.*
Warp filament non-spun ...	Present in North America, Oceania, Central Africa.
Weft filament non-spun ...	Present in N. America, Cent. Africa, Indo-China.
Back-strap	Present in Eastern Asia and Archipelago, Oceania, America.
Spool or bobbin	Present in various stages of form *A* in Asia, America, Europe, and Africa.
Warp-spacer (Fig. 8) ...	Present as reed in Asia, Europe, and parts of Africa.
Sword or beater-in ...	Doubtful similarity in North and South America.
Heddle-rod	Same distribution as back-strap, also in Africa in places where cotton weaving has been introduced without the reed.
Heddle-rod frame (Fig. 21)	Absent elsewhere.
Shed-stick	Present in some forms elsewhere.
Shed-stick and warp-spacer (Figs. 10 and 12).	Absent elsewhere, except perhaps in America.
Warp-peg without warp-beam	Present in Africa.
Laze-rods *absent*	Absent (?) in America occasionally.

The question arises. Did the Ainu invent any of the above loom constituents apparently peculiarly their own ? To enable us to form any opinion, we must get some notion of their capacity for development or invention, which at a time when they have been driven into cold northerly regions and more or less completely cut off from any outside stimulus except that of their conquerors, is somewhat difficult to do. Judging from the statements made by eye-witnesses and students, the Ainu do not show any capacity for improvement. Von Brandt, German Consul-General for Japan, writing forty years ago on the contact between the Ainu and Japanese, says :—" The Ainos, in spite of this contact, continuing for thousands of years, have adopted nothing from the Japanese ; they are what they were—a race standing at the lowest stage of culture, and probably also not capable of

---

[1] In the first edition of his book the names of two of the constituents have in error been transposed.

civilization. . . . ."[1] A very sweeping assertion; nor have we any means of proving that the contact has continued for thousands of years, and it is obvious to any student that the Tasmanians, Australians, Fuegians and Punans stood or stand on a lower stage than the Ainu.

A more recent investigator, Romyn Hitchcock, already quoted, is almost equally severe, saying :—" The Aino in close touch with Japanese civilisation remains, intellectually and otherwise, as much a savage in culture to-day as he ever could have been . . . They now use Japanese knives instead of stone implements and metal arrows in place of flints. But it is scarcely a century since they emerged, and otherwise they have not passed beyond it . . . . The Aino has not so much as learned to make a reputable bow and arrow, although in the past he has had to meet the Japanese, who are famous archers, in many battles " (p. 433). As to the bow and arrow, we meet with a flat contradiction from the pen of B. Douglas Howard, according to whom the Ainu are good shots and make a good bow, which at about forty to sixty feet range could be almost as effective as a rifle.[2] This was in Sakhalien, where the Ainu are free from Japanese oppression. It is going a long way beyond our experience of the evolution of culture to expect that a Stone Age people coming into hostile contact with the much higher metal-using civiliza-tion of a more fecund race should adopt some of the latter's culture, especially when, as in the case before us, the Japanese have until recent years been quite oblivious to the interesting character of the Ainu, and have treated them accord-ingly—in other words, oppressed them rigorously.[3] The Ainu are flesh-eaters, but the Japanese do not allow them to kill the native deer, and have taken their fish stations away from them, forcing them to become vegetarians.[4] Such treatment must tend to degeneration, yet, in spite of it, the Ainu have adopted some tools and methods from their oppressors.

Batchelor informs us the Ainu now use Japanese matches instead of obtaining fire by friction with elm roots, and later with flint, and flint and steel, also that they use Japanese razors instead of sharp flints and shells for shaving purposes (*op. cit.*, pp. 47, 139, 149), while Hitchcock has shown that they have discarded flint arrow heads for Japanese steel ones. The Ainu have also introduced Japanese warp threads into their looms. These adoptions apparently needed little mental effort, but judged by their stage of culture, greater than we can perhaps conceive. It is progress in a slow way. They have gone a big step further, for, as A. S. Bickmore recorded some fifty years ago, they had begun to *work* iron, a very remarkable action for a Stone Age people.[5] The advance so made is still due to contact, but it argues ability for improvement in that they understand there are

---

[1] *Journ. Anthrop. Inst.*, iii, 1874, p. 132.
[2] *Life with Trans-Siberian Savages*, London, 1893, p. 80.
[3] It will be understood, of course, that I am speaking of the *past*.
[4] Batchelor, 1901 ed., pp. 17-18.
[5] " Some Notes on the Ainu," *Trans. Ethn. Soc.*, vii, N.S., 1869, p. 17.

warp superior to their own, and are prepared to make an effort to attain the new object.

Further light may possibly be gathered from an examination of their cranial capacity. From the measurements of seven Ainu skulls in the Museum of the Royal College of Surgeons, London, kindly supplied to me by Professor A. Keith, F.R.S., it appears that the average content is 1509 c.c., with a variation from 1400 to 1630 c.c., results which Professor Keith informs me are somewhat on a par with those of the average European. Most students who have gone into the question of the relation of the size of the brain to its intelligence would probably agree that in bulk the size of the brain is an index to at least potential mental ability. From this we may conclude that the Ainu may be quite capable of improvement.

The persistence in the use of the non-spun bast filament, the presence of varied forms of shed-stick, and the absence of laze-rods point to isolation, but the varied forms of the shed-stick also point to progress.

If to the above points tending to show that there is a potentiality for progress and to the actual record of progress we add the otherwise doubtful negative evidence that certain constituents of the Ainu loom are not found elsewhere, we may, I think, acknowledge that these constituents are indigenous to the Ainu, and not due to contact.

4.—SOME AMERICAN LOOMS.

1. In the Royal Scottish Museum, Edinburgh, there is a specimen of a loom obtained from the North American Slave Indians with a porcupine quill fillet in the process of making. In a previous paper[1] I described some of the methods employed by the North American Indians in the production of their quill-work decoration, but the method of manufacture of this fillet is quite different. It does not appear to have been described so far, and seems to me to be worth calling attention to (Figs. 29 and 30).

The frame is merely a piece of a branch about 1 inch (or 2·5 cm.) in diameter, bent artificially into the shape of a bow, the chord being 23½ inches (or 60 cm.) long from tip to tip, with a depth of about 2¾ inches (or 7 cm.). There is a piece of folded, tanned (?) leather fixed at about 3 inches (or 7·5 cm.) distance from one end of the bow, being held in position on the one side by a tie of soft leather (buckskin), and on the other side by a set of twenty-eight pseudo-warp threads. The leather as folded measures 1⅞ × 1¼ inches (or 4·8 × 3·2 cm.). As will be seen directly, these apparently warp threads are only warp-thread supporters. All the threads consist of two lengths of non-coloured sinew twined together. One end of every thread is made fast at the warp end of the bow, passes through the pseudo-warp spacer into the inside of the folded leather by means of a slit at the folding, passes

---

[1] " Moccasins and their Quill Work," *Journ. Roy. Anthrop. Inst.*, xxxviii, 1908, pp. 47-57.

through the loop of the buckskin tie, and returns through the adjoining slit to the bow end it started from.

The pseudo-warps are kept in workable position by a pseudo-warp-spacer, which consists of a piece of birch bark 2 inches wide by $1\frac{1}{4}$ inches wide ($5 \cdot 1$ x $3 \cdot 2$ cm.), perforated with twenty-eight holes in the same horizontal line, the pseudo-warps passing through these holes. When all the pseudo-warps are in position the folded piece of leather is sewn up with a few sinew stitches.

The first traverse, which is only apparently the weft, consists of two pieces of red-stained sinew which are twined alternately under and over the pseudo-warps. Then the sinews from the spool, which is continuous and not stained in any

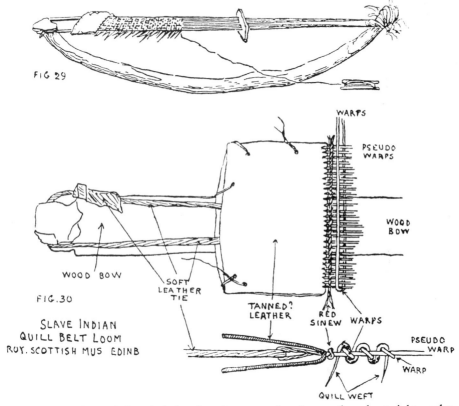

FIG 29

FIG. 30

SLAVE INDIAN
QUILL BELT LOOM
ROY. SCOTTISH MUS EDINB

WOOD BOW

SOFT
LEATHER
TIE

TANNED?
LEATHER

WARPS

PSEUDO
WARPS

WOOD
BOW

RED
SINEW   WARPS

PSEUDO
WARP

WARP

QUILL WEFT

way, is wound round the whole lot, forming a set of real warp threads at right angles to the pseudo-warps, both above and below them.

Variously coloured porcupine quills, which form the weft, are then inserted from below between the pseudo-warps, and bent into position over and under the warp, and so the fabric is made. By well pressing back the warp after the quill insertion, the upper and lower warp are brought into the same vertical plane and remain unseen. It is a very ingenious piece of work.

2. In the Cambridge Museum of Archæology and Ethnology there is a Kachiquel Indian loom brought over in 1885 by A. P. Maudsley. The interesting part about

it is that, after a start has been made at weaving at one of the beams for a length
of 5 inches (or 13 cm.), a second start has been made at the other beam, which
extends to a length of 20½ inches (or 52 cm.); then there are the bare warps
between the two webs for a distance of 54½ inches (or 1·37 m.). From the second
start the weaving would be continued until the two webs meet, where, owing to the
difficulty of making a shed in the ever-narrowing space between the webs, the full
quantity of picks could not be made, and hence there remains a coarseness or open-
ness which is easily noticed. A piece of cloth so woven by the Hopi has been given
to Bankfield Museum by Miss B. Freire-Marreco.

This method of a double start, which may be a substitute for laze-rods, appears
to be an American characteristic, and is not modern, for it shows itself in the upper

FIG. 39.

NAHUA (ANCIENT MEXICAN) GIRL
WEAVING. MENDOZA CODEX, KINGS-
BOROUGH VOL. I. PL. 61. – NOTE THE BIT
OF WEAVING AT THE WARP END A.

portion of the illustration (Fig. 39) of a
loom in the Mendoza Codex as reproduced
by Kingsborough. The dimensions of this
Kachiquel loom are as follows : length from
beam to beam inclusive, 78¾ inches (or
2 m.); the beams are of hard wood, 2·4 cm.
(or about 1 inch) in diameter ; there are 84
warp threads to the inch (or 33 to the cm.);
the warp is woven double (i.e., " sisters," two
threads as one, but *not* of " doubled "[1] yarn);
there are 30 picks to the inch (or 11·8 to
the cm.); the weft is single except for about 14 picks at the heading and tailing.
The temple (Fig. 36), or instrument for keeping the width of the web correct and the
selvedges parallel, is made of a portion of a reed with a piece of needle stuck in
at each end for fixing to the cloth. The temple is placed *underneath* the finished
cloth. The spool is the primitive longitudinal type *A* of my diagram. The
beater-in has a hard convex surface, tapers at both ends with irregular edges.
The shed-stick is of cane, with Balfourian ornamentation at the node, gummed up
at the end, and apparently filled with seed (?), which rattles when shaken. The
" Oxford check " pattern on the cloth is obtained by means of red warp threads at
intervals of 3·5 cm., crossed by red picks at intervals of 4 cm., for which a special
red-yarn spool is provided.

3. A loom from Uitoto, in the Peruvian part of the head of the Amazon district,
and now in the British Museum, has two peculiarities worth mentioning. The
heddle leashes, which are of the spiral form, instead of hanging direct from the
heddle rod, hang from an attached cord (Fig. 34), and the temple (Fig. 37) consists
of a piece of hollow cane with a loose very thin piece of cane running through it,
the protruding ends of which are stuck into the finished portion of the web, practically similar to that of the above-mentioned Kachiquel loom. The dimensions are :

[1] " Doubled " is a term used to denote two or more threads twisted into one, and known
as two-ply, three-ply, six-ply, etc.

length, beam to beam inclusive, about 14 feet (or 4·25 m.) ; width of web, 17½ inches (or 44·5 cm.); 64 warps to the inch (or 25.6 to the cm.); the warps are sisters, same as the Kachiquel warp ; 25 picks to the inch (or 10 to the cm.); the wefts are single.   Both warp and weft are continuous.   The spool is of the primitive longitudinal type (*Aa*).   The shed-stick is a palm midrib or stem.

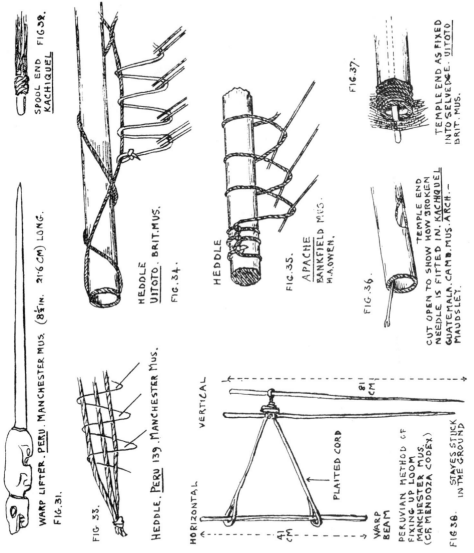

FIG. 32. SPOOL END KACHIQUEL

HEDDLE UITOTO. BRIT.MUS. FIG.34.

FIG.35. APACHE BANKFIELD MUS. M.A.OWEN.

FIG.37. TEMPLE END AS FIXED INTO SELVEDGE. UITOTO BRIT. MUS.

HEDDLE

WARP LIFTER. PERU. MANCHESTER MUS. (8½IN. 21·6 CM) LONG. FIG.31.

FIG.36. CUT OPEN TO SHOW HOW BROKEN NEEDLE IS FITTED IN. KACHIQUEL TEMPLE END GUATEMALA. CAMB.MUS. ARCH.— MAUDSLEY.

FIG 33. HEDDLE. PERU 139. MANCHESTER MUS.

VERTICAL

HORIZONTAL

PLAITED CORD

18 CM

41 CM

WARP BEAM

STAVES STUCK IN THE GROUND

PERUVIAN METHOD OF FIXING UP LOOM MANCHESTER MUS. (CF MENDOZA CODEX) FIG 30.

4. A very interesting loom (Plate I) is that brought from Mazatec, Arizona, by J. Cooper Clark, and now in the British Museum ; for besides the plain up-and-down web, a large portion is devoted to twist or gauze[1] weaving, while a considerable piece of the plain web is afterwards covered by a woven-in design of dark blue

---

[1] Gauze, formed by crossing adjacent warp threads and bound by weft at the point of junction,

wool. Beginning at the breast beam, there are 10 plain picks, then 1 of gauze, then 4 more plain picks, whence the gauze weaving extends a length of about $8\frac{1}{2}$ inches (or 22 cm.), and on this is woven the pattern just mentioned ; then we have a further $2\frac{5}{8}$ inches (or 6·5 cm.) of gauze ; then $\frac{9}{16}$ inch (or 1·4 cm.) of plain weaving, followed by 3 inches (or 7·6 cm.) of gauze, and so on. The warp lay-out for accomplishing the gauze is shown in Fig. 41. It should be noted that on this loom, as on the Kachiquel loom, a piece of the tailing has been completed before the heading was started on. The dimensions are : length, beam to beam inclusive

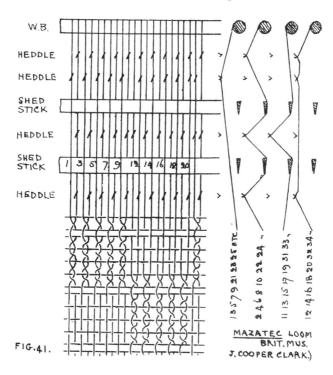

FIG. 41.

MAZATEC LOOM
BRIT. MUS.
J. COOPER CLARK.)

92 inches (or 2·35 m.); width of web, 19 inches (or 48·25 cm.). There are about 42 warp threads to the inch (or 16.5 to the cm·), and 24 picks of two threads each (*not* " doubled ") to the inch (or 11·6 to the cm.). There are four heddles, the rods of which are 11, 9, 7 and 8 mm. respectively in diameter. With so many heddles, laze-rods may not be necessary, but amongst the loose sticks with the loom some may have served as laze-rods. Two fish-ribs are stuck into the cloth, probably for picking up missed threads. Form of spool is longitudinal, corresponding to form *Aa*. Crawford[1] mentions that gauze-weaving is common among Peruvian textiles.

5. In the Manchester Museum there is an ancient Peruvian loom (marked " No. 139, Dr. Smithies ") which calls for attention, as it exemplifies a method of pattern weaving found also in Africa, to which I will refer later on, and to a lesser

[1] " Peruvian Textiles," by M. D. C. Crawford (p. 98), *Anthrop. Papers, Amer. Mus. Nat. Hist.*, xii, pt. iii, New York, 1915.

extent in, so far as I know, the Eastern Archipelago.   This method consists in preparing the pattern in the warp so that the weaver not only has the pattern in front of him, but is also, by the arrangement of the warps, guided as to where the pattern is to be placed, and so ensuring regularity.

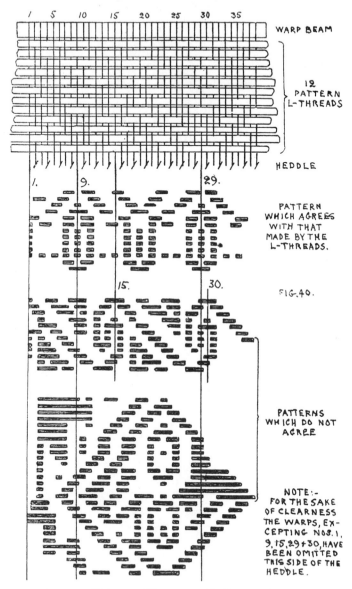

ANCIENT PERU. No 139. MANCHESTER MUSEUM

As will be seen from the diagram (Fig. 40), there are twelve pattern laze threads interlaced in the warp ; by this regulated interlacing the pattern can be distinguished, and it is clearly reproduced in the web, in so far as that has been completed. It is probable that, as the weaving progressed, the pattern laze threads nearest the web, and therefore done with, would be removed, and if necessary

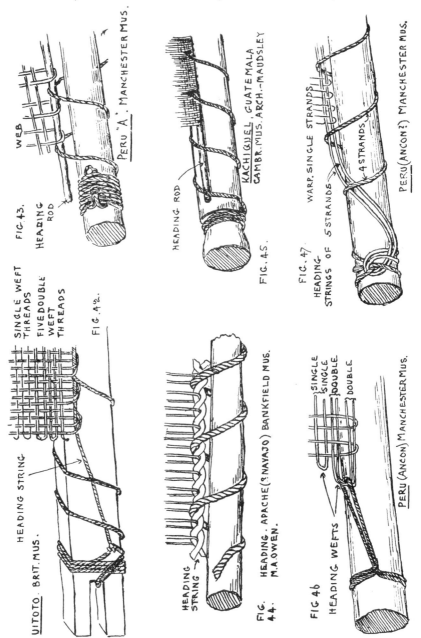

FIG. 43.

WEB

HEADING ROD

PERU "A". MANCHESTER MUS.

SINGLE WEFT THREADS

FIVE DOUBLE WEFT THREADS

FIG. 42.

HEADING STRING

UITOTO. BRIT. MUS.

HEADING ROD

KACHIGUEL, GUATEMALA CAMBR. MUS. ARCH.—MAUDSLEY

FIG. 45.

HEADING STRING

HEADING. APACHE (? NAVAJO) BANKFIELD MUS. M. A. OWEN.

FIG. 44.

WARP. SINGLE STRANDS

STRINGS OF 5 STRANDS

4 STRANDS

FIG. 47. HEADING

PERU (ANCON?) MANCHESTER MUS.

SINGLE
SINGLE
DOUBLE

FIG 46 HEADING WEFTS

PERU (ANCON) MANCHESTER MUS.

re-inserted above the others in such new arrangement as might have been necessary. It will be noticed that the pattern commences to repeat itself at every twentieth warp. Not so, however, as regards the second pattern in the web, which repeats itself at every fifteenth warp ; hence the arrangement shown in the warp does not apply to the second one, and hence also it is obvious that for every different pattern there must be a different arrangement of the laze threads interlacing the warp. There is a third pattern which does not agree with either the first or the second, and there is also a fourth (in tapestry work, but not shown on the diagram), which likewise does not agree with the others. Crawford speaks of short pieces of cane found amongst the *debris* of Peruvian tombs and also on unfinished Peruvian looms which apparently serve the same purpose as these pattern laze threads. The heddle consists of three pieces of cord (Fig. 33) without any rod, nor is there anything to indicate that there ever was a rod.

The dimensions of this loom are : length, beam to beam inclusive, 22½ inches (or 57 cm.); width of web, 16½ inches (or 42 cm.) ; about 19 warps to the inch (or 7·5 to the cm.); 144 picks to the inch (or 57 to the cm.) in the plain web and 93 to the inch (or 37 to the cm.) in the pattern.

Among some loose weaving and spinning tools from ancient Peru in the Manchester Museum are some staves (Fig. 38), evidently for holding the loom in position while weaving is in progress, the warp end being fixed on to the staff and the breast beam being held by means of a waistband round the body of the kneeling Nahua weaver, as shown in the Mendoza Codex (Fig. 39). On one loom, marked " Apache " (? Navajo) in Bankfield Museum the heddle leashes are crossed spiral in form (Fig. 35).

There is in all these looms a very great diversity in the warp attachments to the breast beams or headings, as shown in Figs. 42—47. The warps are attached to a heading string or rod, which in turn is attached to the breast beam by a binding string. In one specimen (Fig. 44) the two heading cords are twisted so as to catch a warp loop at every twist, and so act as a warp spacer or laze rod, while in another (Fig. 40) the web commences without any heading string or rod at all. In none of them is the warp supported directly on the beam. Whether this is so in American looms in general I cannot say.

5. AFRICAN LOOMS.

IN so far as my information extends there are seven forms of looms in Africa, with local variations, which, considering the enormous area of that continent, its great population with its ceaseless migrations may, perhaps, not be considered much, yet in this respect it appears to be more prolific than either the Asiatic or American Continents. The forms are :—

1. The Vertical Mat Loom.
2. The Horizontal Fixed Heddle Loom.
3. The Vertical Cotton Loom.
4. The Horizontal Narrow Band Treadle Loom.
5. The Pit Treadle Loom.
6. The Mediterranean or Asiatic Treadle Loom.
7. The " Carton " Loom.

These forms are easily distinguishable and occupy distinct areas, although in parts they overlap considerably.

1. *The Vertical Mat Loom.*—This loom, the most primitive of all, has a wide distribution, extending from the West Coast to the east of the Great Congo Basin, and is often spoken of as a grass loom on account of the warp and weft (neither of which is twisted or spun) having the appearance of grass. The filament used is, however, obtained from the leaves of the Raphia palms, *Raphia ruffa*, *Mast.* and *R. vinifera*, which flourish, the former in East Africa and Madagascar, and the latter in West Africa. The outer cuticle of the leaf is drawn off and the underpart cut into thin filaments by means of a leaf splitter, Figs. 48A and B. The specimen in Bankfield Museum consists of 109 thin slips of cane, 4 mm. wide, securely and ingeniously fastened together and fitted into a suitable frame. The loose ends of the slips of cane are pointed, and when the splitter is drawn lengthwise along the surface of the flayed cuticle it cuts it up into numerous filaments which are used as warp and weft without further preparation. Besides the raphia leaf filament, Sir H. H. Johnston[1] informs us that in the western and south-western Congo basin short cloths were also made from grass.

---

[1] *Geo. Grenfell and the Congo* (London, 1908), ii, p. 589, footnote.

26

The loom has two representatives in Bankfield Museum, one from the Kwa Ibo River, West Africa, kindly given to the Museum by the late Mr. John Holt, a well-known Liverpool merchant, in 1900, and the other from the Ba-Pindi people, in Central Congo, obtained in 1909 through the kindness of Mr. E. Torday.

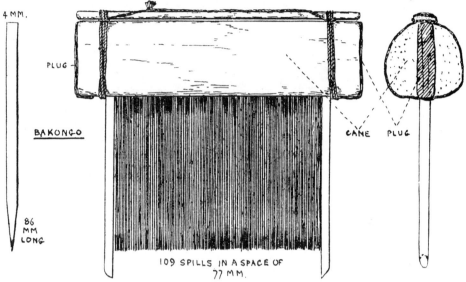

4 MM.

PLUG

BAKONGO

CANE   PLUG

86 MM LONG

109 SPILLS IN A SPACE OF 77 MM.

FIG 48A. BAKONGO LEAF SPLITTER BANKFIELD MUSEUM (E.TORDAY)

The Kwa Ibo loom is evidently a very close facsimile of the one depicted by Du Chaillu as in use by the Ishogo, Fig. 49. The web, or woven mat, width is approximately 16 inches (or 41 cm.), and its length from beam to beam is about 57 inches (or 1·45 m.). The warp beam consists of a piece of tree branch without the bark, 32 inches (or 81 cm.), long. The breast beam consists of a portion of palm leaf mid-rib or stem, common to all these looms, having a large slot at either end wherewith to fix it on to its upright supports.

The method of attaching the warp to the breast and warp-beams (see Figs. 50 and 51A and B) is, as in all these looms, a complicated one, on account, no doubt, of the comparative smoothness of the filament, which does not bind well. The warp filaments are split up into seventy-three bunches, and their ends knotted on to a heading rod which is fitted into the groove of the breast beam, Fig. 50, all being held in position by some twisted lashing.

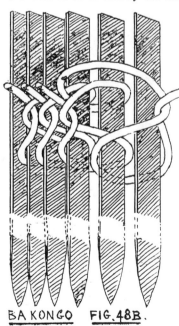

BAKONGO   FIG.48B.

At a distance of about 45 inches (or 1·14 m.), away from the breast beam, the warps are again bunched, but this time into fifty-seven bunches, a number which naturally does not agree with the bunching at the breast beam. These bunches are attached to the warp beam by intermediary cords, with slip knots, Fig. 58, which are wound twice round the warp beam and then, accumulating as they proceed from left to right, run along it towards the top right-hand corner, where they are tied into one big knot.

ISHOGO MAN WEAVING . FROM
DU CHAILLU'S ASHANGO-LAND. LONDON.1867.
FIG. 49.

The heddle, according to Du Chaillu's drawing, Fig. 49, looks as though it were in reality two heddles, and Ephraim has taken it to be such.[1]   But there is only one heddle, Fig. 51A, the explanation being that the heddle rod consists of two independent parts which, for the sake of convenience in weaving, the worker holds apart with his fingers and thumb and so misleads one at a cursory glance.   Both parts,

[1] *Op. cit.*, p. 18.

28

by the way, are made up of two pieces of split cane, Fig. 51B, but that does not affect the question. A varying quantity of filaments is made up like a skein, knotted at certain intervals and placed zig-zag between the higher pair of split canes and the lower pair, Fig, 51A, and fastened in position so that the knots appear just above where the split canes are tied together, Fig. 51B. In this heddle there

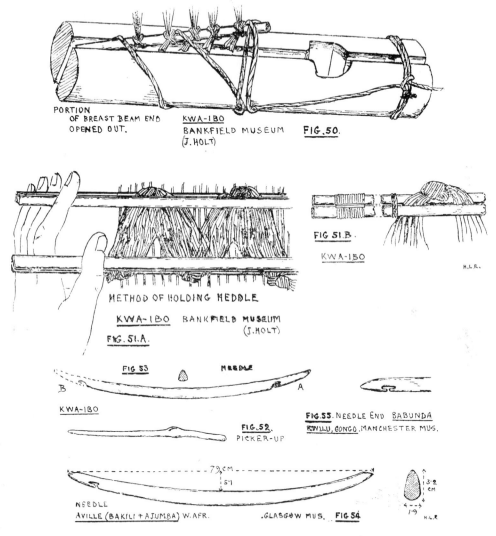

PORTION
OF BREAST BEAM END
OPENED OUT.   KWA-IBO
          BANKFIELD MUSEUM   FIG.50.
          (J. HOLT)

FIG 51.B.

KWA-IBO

H.L.R.

METHOD OF HOLDING HEDDLE

KWA-IBO   BANKFIELD MUSEUM
                    (J.HOLT)
FIG. 51.A.

FIG 53   NEEDLE

B   A

KWA-IBO

FIG.55. NEEDLE END BABUNDA
KWILU, CONGO. MANCHESTER MUS.

FIG.52.
PICKER-UP

79 cm

5·1

NEEDLE
AVILLE (BAKILI + AJUMBA) W. AFR.   .GLASGOW MUS.   FIG 54.

3·2
cm

1·9   H.L.R.

are eight such skeins, and generally speaking some of the filaments of the adjacent skeins are made continuous, but with others they are not so. Some of the individual warp filaments are held up to the heddle rod by three leash filaments, some by as many as ten—there is no fixed rule—the irregularity being apparently due to the irregular splitting of the leaf. In working, as shown, the warp between the two sets of rods is barely visible, being covered up by the profusion of leash filaments.

29

The loom is provided with two laze rods 14 and 16 inches (or 36 and 41 cm.), long respectively and ⅝ and ¾ inch (or 1·6 and 1·9 cm.), in diameter ; one rod is therefore shorter than the width of the web and the other only just a little longer. It is provided with a picker or warp raiser, Fig. 52, which is nothing more than a smoothed branch 15 inches (or 38 cm.), long, tapering to a blunt point at one end.

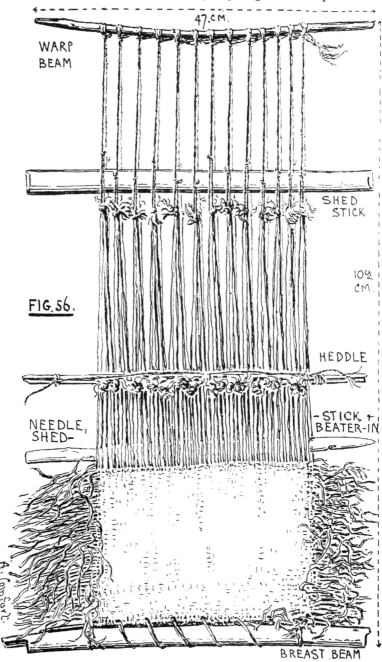

47·CM.

WARP
BEAM

SHED
STICK

102
CM.

FIG.56.

HEDDLE

-STICK +
BEATER-IN

NEEDLE,
SHED-

BREAST BEAM

30   BA-PINDI LOOM BANKFIELD MUSEUM (E.TORDAY)

In all these looms the sword consists of a combination of needle, shed stick, and beater-in combined, Fig. 53. It is of a hard dark wood, somewhat curved longitudinally and is, as an exception, furnished with two nicks for carrying the weft.

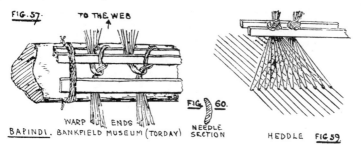

FIG. 57. TO THE WEB

WARP ENDS

BA-PINDI. BANKFIELD MUSEUM (TORDAY)

FIG. 60. NEEDLE SECTION

HEDDLE FIG. 59

The nicks are invariably cut towards the adjacent needle point so that, as the point of the sword is used as a shed opener, it would seem the method of working is to put the sword or needle through the shed, fit the weft into the notch and draw back the sword, which draws the weft with it and makes the pick. This method does not agree with the details of Du Chaillu's illustration, Fig. 49. In a Babunda needle in the Manchester Museum the nick is cut both ways, Fig. 55.

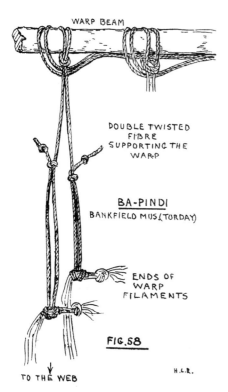

WARP BEAM

DOUBLE TWISTED FIBRE SUPPORTING THE WARP

BA-PINDI BANKFIELD MUS. (TORDAY)

ENDS OF WARP FILAMENTS

FIG. 58

TO THE WEB

H.L.R.

The weft is discontinuous, each piece being a few inches longer than the width of the web; there is no selvedge, and hence no temple is used. There are 22 picks to the inch (or 8·7 to the cm.), and 31 warps to the inch (or 12·3 to the cm.).

The Ba-Pindi loom already referred to, Fig. 56, differs from the Kwa Ibo loom in some details worth noting. The length overall is 49 inches (or 1·24 m.), with a web width of 18½ inches (or 47 cm.). The method of attaching the warp ends to the breast beam is as follows, Fig. 57. The ends are passed between a pair of thin heading rods, then over and under both pieces and into the loop thus formed, which on drawing tight becomes a knot. This arrangement is placed on the beam and an extra batten, consisting of a wider and larger piece of cane, placed over the warp ends to just below the knots and then all lashed on to the beam by means of some coarse cord. In another specimen in Bankfield Musuem this extra batten is cut out of the breast beam itself and fitted back into its place with the warp ends between it and the beam, which ensures a firmer grip. About 30 inches (or 76 cm.), more or less, from the breast beam the warps are bunched together into twenty-six lots and connected up with the warp beam as in the Kwa Ibo loom, only, instead of all the cord ends being carried to the right-hand top end and there tied into a big knot, they are cut off at various lengths.

The heddle, Fig. 59, consists of two strips of cane between and on which rest eight sets of two knots each of the leash ends which support the warp. Every leash, like the warp and weft, consists of about twelve to fifteen separate filaments. Each set of leashes is distinct from the next, *i.e.*, is not continuous and extends only from one knot to the other of its set and not beyond, and is so arranged that when the knots are placed side by side the leashes separate out and cross one another.

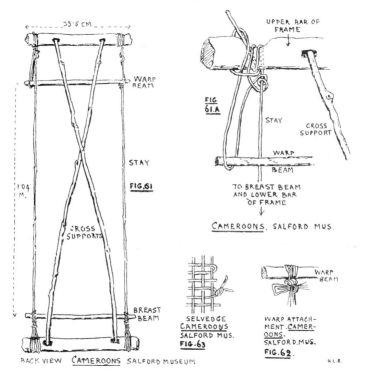

BACK VIEW CAMEROONS SALFORD MUSEUM

There is one shed stick made of palm leaf mid-rib. The needle is curved in transverse section, Fig. 60, with the working- or beating-in edge almost as thick as the back or opposite edge, which is usually broader. Sometimes both edges are sharp and frequently the working-edge is serrated with wear.

In a loom from the Cameroons in the Royal Salford Art Gallery and Museum there is an arrangement, Figs. 61 and 61A, found also elsewhere in West Africa, for obtaining rigidity in the loom frame and therefore better weaving. It consists of two stout rods held apart by means of two cross supports (wood branches) and triced together by means of stays (lianas), the breast and warp beams being made fast to the stays. The holes in the two stout rods into which the cross supports are fitted are at the back of these rods and the cross supports are curved like a bent

32

bow and so act like a spring in keeping the stout rods and beams well apart. The warp attachment is very simple, Fig. 62.

On this loom the web (which is omitted from the illustration for the sake of clearness) shows an incipient stage of selvedge. Occasionally longer pieces of weft than merely suffice for one pick are used and are turned back at the edge ready for the next pick, so making a selvedge, Fig. 63, or occasionally two weft ends are tied together in a knot at the edge, which again forms a selvedge. The casual, and therefore early stage, of the selvedge is indicated by the fact that these knots are 18, 8, 16, etc., picks apart ; neither do they correspond at the opposite edge. Other details of this loom are : width of web 10 inches (or 25·4 cm.), and about eight picks to the inch, or three to the cm.

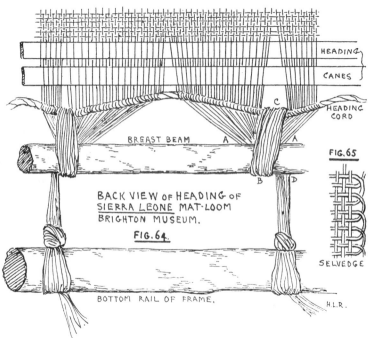

BACK VIEW OF HEADING OF SIERRA LEONE MAT-LOOM BRIGHTON MUSEUM.

FIG. 64.

A complicated form of heading is shown in a loom, said to come from Sierra Leone, in the Brighton Museum, added to the collection there in 1886, and is explained by the illustration, Fig. 64. The details are : frame supported by cross supports and stays is 57¾ inches (or 1·47 m.), long by 27½ inches (or 70 cm.), broad. Length, breast beam to warp beam inclusive, 35·5 inches (or 90 cm.), and width of web 8¼ inches (or 21 cm.). Approximate number of warps to the inch 42½ (or 16·6 to the cm.), and 20⅓ wefts to the inch (or 8 per cm.). The weft is not continuous, but there is a perfect selvedge, Fig. 65. A somewhat similar selvedge is found on an Old Calabar loom in Bankfield Museum, Fig. 75. A longitudinal pattern is obtained by dyeing the warp previous to laying out.

A different form of heading arrangement is shown on a loom, Fig. 66, of unknown provenance in the Royal Scottish Museum, Edinburgh, where the cord

c

fastening the heading rods to the breast beam is put through holes in the latter instead of winding round the beam as is usually the case. In a loom from Mongo in the same collection (Edinburgh) there is complete selvedge, Fig. 67, on both edges, which, however, is not brought about by a continuous weft. After the pick is made the end of the weft is returned over the pick for a distance of about half an inch (or 11 mm.), when it is allowed to emerge, and floats free like a sort of inner fringe. In this loom the warp attachments are simple, Fig. 68, and the sword while curved in transverse section is longitudinally quite straight, which appears exceptional, Fig. 69.

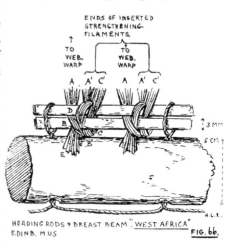

HEADING RODS & BREAST BEAM "WEST AFRICA"
EDINB. MUS.
FIG. 66.

The Liverpool Museum possesses a bag loom, Fig. 70, with completely woven bag still in position, and the Glasgow Art Gallery and Museum possesses another such loom on which only a portion of one side of the bag loom has been completed. As the incomplete bag makes a more interesting study, it will be as well to describe the Glasgow one. The frame is supported by cross supports and stays as in Fig. 61. The warp is so beamed that one side of the bag may be woven first, then when completed the loom is turned back to front and the other side of the bag woven. The method of division of the warp for the front and back is shown in Fig. 74 : it is the same for both looms. Curiously enough there are nearly double the number of warp in the set for the back, yet to be commenced upon, than there are in the set for the half-finished front. Perhaps some of the former are cut away when weaving commences, or perhaps they are preparatory for the two sides of a second bag off the same loom with the same beaming. The shape of the bags is that of a

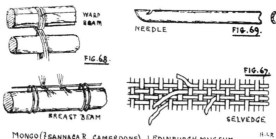

MONGO (?SANNAGA R. CAMEROONS). | EDINBURGH MUSEUM.

truncated isosceles triangle, with top and bottom parallel but the bottom narrower than the top, with the sides expanding regularly from bottom to top.

On the half-finished front side there are 62 bunches of warp covering a width of 18½ inches (or 47 c.m.) on the warp beam ; these bunches are reduced in number to 21 thicker bunches covering a width of 8 inches (or 20 cm.), on the breast beam. There is no inserted warp, the number of filaments on both beams being the same ; the number of warps *to the inch* is therefore more on the breast beam than on the

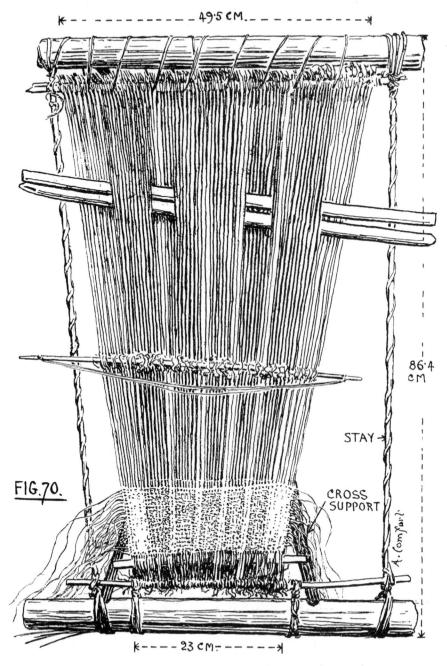

FIG. 70.

49·5 CM.

86·4 CM

STAY →

CROSS SUPPORT

23 CM.

OLD CALABAR, LIVERPOOL MUS.   BAG LOOM (RIDYARD).

warp beam, being compressed into 10½ inches (or 27 cm.) *less* space than on the warp beam. The bunches are :

Warp beam : 17 non-coloured, 8 black, 12 red, 8 black, 17 non-coloured.

Breast beam : 7 non-coloured, 2 black, 4 red, 2 black, 6 non-coloured.

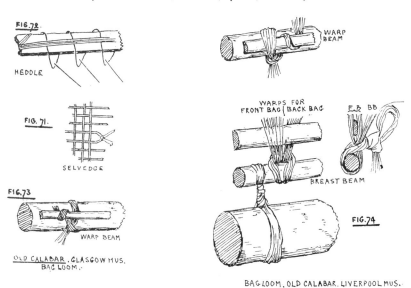

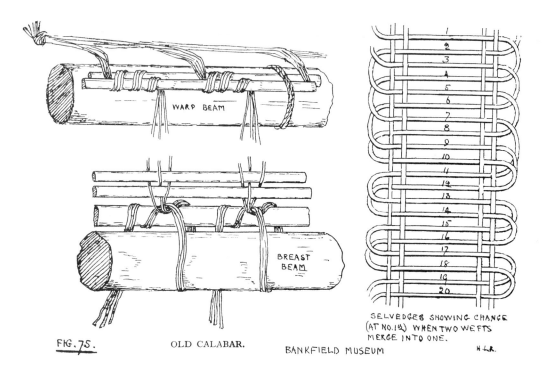

FIG. 75.   OLD CALABAR.

BANKFIELD MUSEUM

SELVEDGES SHOWING CHANGE (AT NO. 12) WHEN TWO WEFTS MERGE INTO ONE.

To prevent the outer warp getting awry, at intervals of $1\frac{1}{2}$-2 inches (or 4-5 cm.), the weft ends are knotted together over the outermost warp, Fig. 71 ; but this can only be a temporary or working selvedge to be undone preparatory to interlacing the finished woven front and back.   On the back, evidently to keep the warp of the bag from getting entangled with those of the

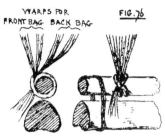

WARPS FOR FIG.76
FRONT BAG BACK BAG

BANANA . EDINBURGH MUSEUM.

front, seven laze threads of twisted fibre have been drawn irregularly through it, the ends of these laze threads being fastened to the stays.

The heddle rod is a flat piece of wood having the leashes kept in position by means of longitudinal cords.   Details of the warp attachments are given in Figs. 73 and 74.

In connection with bag looms may be noticed one from Banana, Congo River, in the Royal Scottish Museum, Edinburgh, which is prepared for weaving two bags (or four mats ?) from one and the same breast beam, but with distinct warp beams. Each set of warp is provided with a Du Chaillu heddle, laze rod and needle, the warp ends being fixed along the warp beam in the usual method.   All the needles are concave on the working edge with corresponding convexity on the back edge. The width of the weave is $16\frac{1}{2}$ inches (or 42 cm.), fine work.   The object of weaving four sides off one breast beam will save labour in beaming ; but only one person can work at it at a time, for all the heddles are placed on the same side of the warp and two people working at it would interfere with each other.

A very interesting loom is one marked Okale (Ba-Hamba) in the British Museum.   It is in most respects like the rest of these looms, but shows a pattern (Fig. 77) obtained by means of black-stained wefts, the pattern being roughly arranged in the warp near the warp beam by means of 36 strips of cane 4 mm. wide, which are in fact pattern laze rods.   In this specimen, owing to previous rough handling, I have not been able to prove conclusively the connection

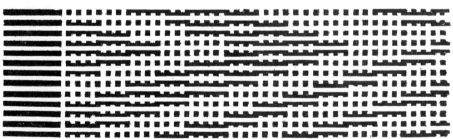

FIG.77. PORTION OF PATTERN ON OKALE (BA-HAMBA) LOOM, BRIT. MUS. (TORDAY).

between the two by running the fingers along the warp, but that a connection exists is evident from the illustration.   As already mentioned (p.23), this method is found in Ancient Peru and in the East.   The dimensions of the loom are : length, breast beam to warp beam inclusive, 34 inches (or 86 cm.), the knotted warp ends hanging down a further 12 inches (or 30 cm.) ; width of web 16 inches (or 40·5 cm.) ; 30 warps to the inch (or 11·8 to the cm.) ; 24 picks to the inch (or 9·5 to the cm.). The warp is always in pairs (" sisters ") and hence passes through the leashes in twos ; these leashes are of finer filament than the warp and weft. The needle is of the usual hard wood, slightly concave on the working edge, which is blunt, but without serration. The selvedge is apparently made after the completion of the weaving, but there are selvedge knots, as shown in Fig. 71, every 20 to 24 picks. Details of the complicated warp attachment to the breast beam are given in Fig. 78.

As to the origin of this mat loom there is no other loom in any way comparable with it except perhaps the Vertical Cotton Loom discussed on pp. 49-58, and when we have said that they are both upright looms and are furnished with a heddle, the comparison is at an end. There is a great gulf between this mat

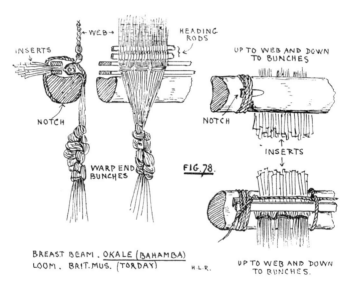

BREAST BEAM . OKALE (BAHAMBA)
LOOM . BRIT.MUS. (TORDAY)   H.L.R.

loom and the Ancient Egyptian vertical looms, for the illustrations of which we are indebted to N. de G. Davies.[1] Both are upright, both have a heddle, and both are worked by men—as a rule. The Ancient Egyptian weaver used a ball of yarn for his weft, while the modern African uses a needle as weft-carrier, which serves also as a shed stick and beater-in. The Ancient Egyptian loom had, in so far as we can judge, ordinary heddle leashes, which were not bunched, and the African weavers have bunched leashes. The only comparison one can make is with the bunching of the leashes on the Livlezi loom, Fig. 87. This, however, gives one the impression of a raphia weaver adopting his own method with an introduced filament, i.e. cotton, and as the bunching lessens the control of the weaver over his warp there is not the likelihood that the cotton weaver adopted the raphia weaver's method.

[1] See *Ancient Egyptian and Greek Looms*, by H. Ling Roth (Halifax, 1913).

38

All the intact specimens which have come under my observation show considerable neatness in the make, being well and carefully put together ; the weft-carrier is nicely finished ; the comb-like leaf cuticle splitter is a trim little article ; the work produced is excellent of its kind, especially the embroidery work of the Ba-Kongo, for instance, which, although considered to be a recent introduction,[1] is fine and artistic. As will have been seen, the selvedges are in various stages of development, and the heddles show some variation in their leash attachments. The parts may be crude, but they are not slovenly made, and it is very clear that much care has been devoted to getting both the loom and the web to a comparatively high pitch of excellence. Altogether one gets the impression that the makers and users of this form of loom are a progressive people. The form is, however, extremely primitive, and this, together with the mat work found side by side, tempts one to conclude that the form may be indigenous to the habitat of the raphia palms. But before adopting such a conclusion it will be as well to examine the various steps apparently necessary to be taken in the transition from mat work[2] to weaving, for the majority of students who have looked into the question of the origin of weaving are of opinion that it originated in basketry or matmaking.

The transition appears to be due to an appreciation of the principle of the heddle, as yet unknown, and the translation of that principle into a mechanical factor. The principle is already in action when, in making a mat, the worker raises (1) one of the filaments, the warp, to pass or interlace the other filament, the weft, and it is intensified when he raises two or more warp filaments together (2) with the purpose of saving labour. In so far as one can judge, this would have been followed by permanent bunching—*i.e.* by means of leashes (3)—which would take the place of the fingers, and is the first mechanical step towards the adoption of the heddle. At this point, if not sooner (I judge from my own experiments), it would probably be found that some arrangement (4) is necessary whereby the warp can be kept more or less taut, the matter depending largely on the nature of the material employed. A further advance would consist in attaching the leashes in bunches to short pieces of wood (5) to enable them to be lifted more easily—a sort of handle, in fact—as can be seen in a belt loom from Iceland in Bankfield Museum, where there are three such sets of warp-raising leash-bunches, each attached to a wooden rod 5 cm. long, by means of which the whole of the required warp is raised at thrice. In the Ba-Pindi loom we have the complete transition where the leashes, although still bunched, connect the warp to a single rod (6), whereby the whole of the required warp is raised at once and the mechanical factor has come into full play.

Accepting this surmise of the progress of the transition as approximately correct, we are in want of evidence as to steps (3) and (4) in the development of this

---

[1] Torday and Joyce, " Les Bushongo," *Ethnographie*, ser. iii, tome ii, fasc. i (Bruxelles, 1910), p. 45.

[2] I am not here referring to plaited mats, but to mats the components of which are interlaced at right angles to each other without the use of a frame.

mat loom, and for want of this I must for the present withhold any definite conclusion as to whether the loom is indigenous where we now find it.

2. *The Horizontal fixed Heddle Loom.*—This loom, Fig. 79, on which in Madagascar both raphia fibre mats and silk cloths are woven, appears to be used in

MANGANJA LOOM. FROM C.&D. LIVINGSTONE'S EXP.N
TO THE ZAMBEZI. LONDON. 1865. P.112.

FIG. 79.

Africa for weaving cotton only. It is laid stretched out close to the ground, nearer to the ground in Madagascar and North Central Africa than in South Central Africa, and is worked with the usual laze rods, spool, and beater-in, its characteristic being the fixed heddle. At first sight such a fixture makes it look somewhat

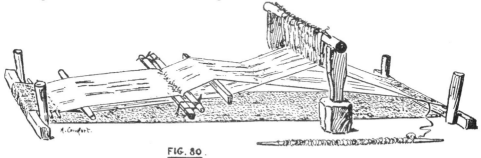

FIG. 80.

WORKING MODEL OF MADAGASCAR MAT LOOM BANKFIELD MUSEUM. (SIBREE)

awkward to work, but on rigging up a similar loom I found I could work it quite comfortably. A good idea of the loom can be obtained from a study of the illustrations, Figs. 80 and 81, Fig. 80 representing a model in Bankfield Museum of a Madagascar mat-weaving loom[1] brought home by Dr. Sibree in 1915, and Fig. 81

[1] Dr. Sibree writes (14th August, 1918) that this is *not* a mat loom, but used for raphia, cotton, hemp and silk.

40

representing a model, likewise in Bankfield Museum, of a loom used by the A-Fipa in their country south-east of the Victoria Nyanza and north-west of Lake Tanganyika, and brought home six years ago by the Rev. Harry Johnson. In the Madagascar loom the warp (? raphia fibre) is continuous, while in the A-Fipa loom in Bankfield Museum, as well as in one from the same people in the Leicester

FIG.81

WORKING MODEL OF A-FIPA COTTON LOOM . BANKFIELD MUSEUM. (H.JOHNSON)

Museum, it is not so. In the Madagascar specimen the weft is likewise continuous, but not so in the A-Fipa loom, where the selvedge is finished in a curious way. The yarn on the spool is, of course, continuous, but when a pick has been made, it appears to have been cut off at both ends about $\frac{1}{2}$ inch (or 1·3 cm.), longer than the width of the web and the over lengths woven in, the result being that the cloth

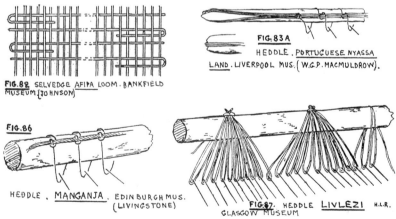

FIG.82 SELVEDGE AFIPA LOOM . BANKFIELD MUSEUM {JOHNSON)

FIG.83A

HEDDLE . PORTUGUESE NYASSA LAND . LIVERPOOL MUS. (W.G.P. MACMULDROW).

FIG.86

HEDDLE . MANGANJA . EDINBURGH MUS. (LIVINGSTONE)

FIG.87. HEDDLE LIVLEZI H.L.R. GLASGOW MUSEUM

for about $\frac{1}{2}$ inch depth for the whole length of both selvedges is much closer than for the rest of the web, as shown in Fig. 82. It reminds one of the selvedge in the mat-weaving loom, as illustrated in Fig. 67, with this difference, that in the A-Fipa cloth the over-length is placed *by the side* of the pick, while in the Mongo mat it is placed *on top* of the pick.

In 1915 Mr. W. G. P. Macmuldrow gave the Liverpool Museum one of these looms from Portuguese Nyassaland, but unfortunately without the frame supporting the heddle. He also gave that Museum a photograph of a native boy weaving, and another of a native boy rigging up the heddle for warp-laying.[1] The photographs, owing to difficulties in the taking, are not quite so clear as could be desired,

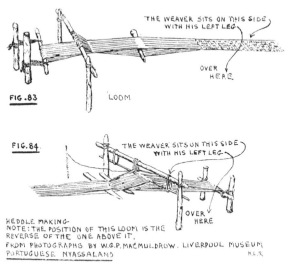

but I think the essentials have been reproduced in the illustrations, Figs. 83 and 84. The dimensions of this Portuguese-Nyassaland loom are : beam to beam inclusive, 67 inches (or 1·70 m.) ; width of web, 4 inches (or 10·2 cm.) ; 25 picks to the inch (or 10 to the cm.) ; length of heddle rod, 25 inches (or 63 cm.), with a diameter of 2 cm.; spool, 22·5 cm. long, of the *Ba* type.   The yarn for both warp and weft does

FIG.85.   Abbild. 23. **Webstuhl vom unteren Sambesi.** Nach Globus 10/1866.

not appear to be indigenous. The warp ends are fixed on to the beams by means of some gluten, which has hardened like dried breadcrumbs. The frame of the model in the Leicester Museum is likewise fixed together with some resinous substance,

---

[1] H. Schurtz, in his *Urgeschichte der Kultur* (Leipzig, 1900), gives us an illustration of a Swahili likewise laying his warp, and has even the heddles in position, but it is somewhat misleading to label as he does, the illustration " Swaheli at a Loom," for the man is *not* weaving.

and in the Bankfield Museum Model the parts are lashed together. In a Livlezi loom, Fig. 87, about to be described, the warp is also fixed to the beams by a sort of resinous gum.

Judging from a photograph of a silk loom placed at my disposal by Dr. Sibree, and partly reproduced in Fig. 90, it would seem that occasionally in Madagascar a second heddle is in use. It is upheld by two iron supports (*BB*), which appear capable of being brought forward towards the weaver and pushed back, actions which must cause the heddle to be lowered and raised.

Dr. Livingstone was the first to give an illustration of the Fixed Heddle Loom. This was in 1865.[1] It is reproduced in Fig. 79. The now defunct journal, *Globus*, reproduced it fairly well (No. X, 1866), but with embellishments, and with the mistake of taking the spool for an ordinary stick. Ephraim, without verifying his quotation, ignores Livingstone in the matter, ascribes the loom to *Globus*, and reproduces it past all recognition, as shown in Fig. 85. Livingstone does not describe it, but he brought home a specimen which he obtained from the Ma-Nganja, south of Lake Nyassa, and which is now in the Royal Scottish Museum, Edinburgh. This specimen is unfortunately incomplete, and does not include the heddle supports. Its details are : the warp (of cotton) is many feet long ; the width of the web at the heading is 23 inches (or 58 cm.), but where the work has ceased it is only 20 inches (or 51 cm. wide), so evidently a temple was not in use. The leashes of the heddle are continuous, and are secured in position by knotting over a cord which runs the length of the rod, Fig. 86. The spool is a piece of split-pointed cane 39 inches (or 99 cm.), long.

Another specimen of this loom is to be found in the Glasgow Art Gallery and Museum, marked " Livlezi loom, south of Lake Nyassa." It is likewise a cotton loom. Length, beam to beam inclusive, 61 inches (or 1·54 m.); width of web at heading, 20½ inches (or 52 cm.), but 18 inches (or 46 cm.), at the last pick when work was suspended, indicating absence of temple. Both beams are of hard cane, and as already mentioned, the warp is gummed to them. There are 52 warps to the inch (or 20·5 to the cm.); and 12 picks to the inch (or 4·7 to the cm.). The warp is not continuous but the weft is. The spool is 31 inches (or 78 cm.), long. As in the Ma-Nganja loom the length of the spool appears out of proportion to the width of the web. The heddle leashes are " sisters " and continuous, and 48 or 50 are bunched over the heddle rod, Fig. 87, as in a specimen of the raphia looms of the Ba-Pindi, Fig. 59, as already mentioned. In the Livlezi loom and in the Ma-Nganja loom, as well as in the working models in Bankfield and the Leicester Museums, the yarn used has a strong twist so that the cloth has the well-known crinkled appearance.

What seems to be the same type of loom and found in Darfur is illustrated by Wilson and Felkin.[2] Dr. Felkin says of it : " The looms are very primitive ;

[1] David and Chas. Livingstone, *Narrative of an Expedition to the Zambesi and its Tributaries*, London, John Murray, 1865, p. 112.

[2] *Uganda*, London, 1882, p. 274.

they are very narrow, and are usually placed under the hedge or a tree low down on the ground, with a hole made underneath to accommodate the weaver's legs." The existence of this pit would lead one to infer that we have to do here with a pit-treadle loom, but the pit is an innovation in connection with this loom, due to contact with the Hindu pit-treadle loom, about which a few words shall be said directly.

It may be mentioned in parentheses that innovations in details, obviously from the East Indies or Arabia, are very common in the lands bordering the East coast of Africa ; thus, for example, Miss Werner illustrates in her work[1] a native loom which, while then in use in that part of Africa, must have come from elsewhere, a good but troublesome example of the migration of the arts.

On the face of it Wilson and Felkin's illustration, although it shows clearly the fixed heddle, is defective, and on my asking Dr. Felkin about it he very kindly replied (May, 1916) and acknowledged the incorrectness of the drawing. It is reproduced in Fig. 88. In writing me, Dr. Felkin mentioned that the weaver

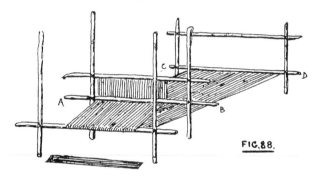

FIG. 88.

LOOM , DARFOUR . WILSON & FELKIN'S UGANDA . LONDON.1868. II .
[ROD AB SHOULD BE OMITTED & WARP FROM AB TO CD INSERTED]

" with a stick of hard wood beats the weft away from him. This is the usual method, but I have seen looms where the weaver beats the weft towards him. This is rare, however. As a foot or so of cloth is finished it should be wound up at the back part of the loom if the weaver beats the weft away from him, the reverse is done if he beats the weft towards him." So that the loom is altogether anomalous, as it shows contact on more sides than one. Up to this point, in so far as my knowledge of these looms goes, there is no winding-up of the web as it gets woven, for in the models in Bankfield and the Leicester Museums there are no movable breast beams, although these are good working models, and in the Madagascar and other illustrations the warp is continuous—like that of a seamless garment—and the warp is shifted round as the weaving goes on, which only ceases when the heading nearly meets the tailing. Incidentally it should be stated that Ephraim, again omitting to verify his quotation, reproduces Felkin's illustration and incorrectly ascribes it to Prof. R. Hartmann, who had reproduced it in a popular work

[1] *The Natives of British Central Africa*, London, 1906 (to face p. 196).

entitled *Die Nillaender*. On the other hand, it may be that Felkin's loom marks a point of contact between the southern and northern distribution of the Fixed Heddle Loom, for further north the warp is much longer and does get wound up.

Fig. 89 shows a reproduction of a water-colour sketch by Frederick Goodall, R.A., in Bankfield Museum, of a weaver at work in Upper Egypt. We must not expect accuracy from an artist, but in this sketch I venture to think that he has indicated very well the fixedness of the heddle which places this loom in the class under discussion, only, instead of being supported from a frame above, or resting on wooden supports, it rests on a couple of stones. Then there is the very broad loom used by the Bedawin in Upper Egypt, and apparently along the whole length of the northern portion of the Sahara, which seems to me to be similar to that depicted in Fig. 89A, and which Franz Stuhlmann[1] considers to be identical with the vertical loom, only laid flat. But to me it appears to be a modification

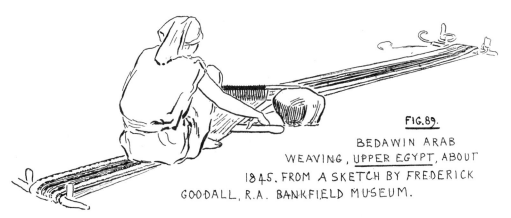

FIG.89.

BEDAWIN ARAB WEAVING, UPPER EGYPT, ABOUT 1845. FROM A SKETCH BY FREDERICK GOODALL, R.A. BANKFIELD MUSEUM.

of the Fixed Heddle Loom. I have *not* seen this loom and only know it by means of illustrations. There is a very poor reproduction of one entitled : *Femmes tissand le Felidj (toile de tente) dans un campement d' Aures*, facing p. 426 of Lieut.-Col. de L'Artigue's *Monograph de l'Aures*, Constantine (Algiers), 1904. There is a somewhat similar one in Madame Jean Pommerol's *Among the Women of the Sahara*, London, 1900, p. 307; one illustrated by R. Karutz in *Globus*, 1907, XCII, No. 8, p. 119, who by the way mentions that on one occasion two women wove a tent cloth on it ; and one by Frederick Goodall, above mentioned, and illustrated in Fig. 89A.

The Ancient Egyptians depicted a loom, pegged out like this one, in the Tombs of Chnem-hotep, but the position of the woman's hand at the end of the heddle and the absence of any indication of a support tends to the view that it is not a fixed heddle. On the other hand, the illustration in the tomb of the Vizier Daga, drawn by N. de G. Davies,[2] shows the weavers' hands quite clear of the

1 *Ausflug in den Aures*, Hamburg, 1912, pp. 116, 118.
2 *Five Theban Tombs*, Plate XXXVII.

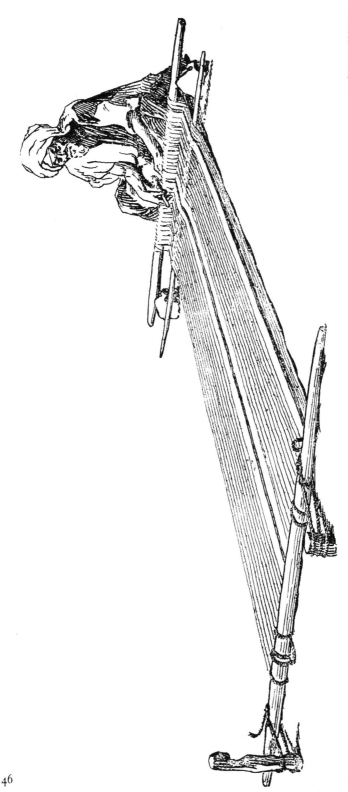

FIG. 89A BEDAWEN WEAVER. FROM THE PAINTING BY FRED K. GOODALL R.A. ENTITLED SPINNERS & WEAVERS. REPRODUCED BY PERMISSION OF THE OWNER, MR W. K. D'ARCY OF STANMORE HALL, STANMORE FROM THE ILLUSTRATION IN CASSELL & CO.'S ROYAL ACADEMY PICTURES, 1892.

heddle, with a curious hook-shaped contrivance at either end, which might possibly be construed into some sort of support, but it is extremely doubtful.   Garstang's

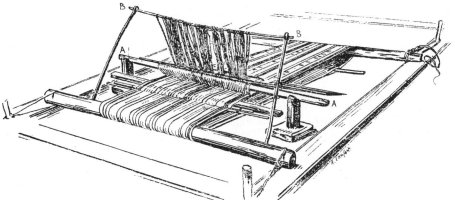

FIG. 90. MADAGASCAR LOOM FROM A PHOTOGRAPH (REV. DR. SIBREE)_ AA FIXED HEDDLE ROD.   BB LOOSE HEDDLE ROD.

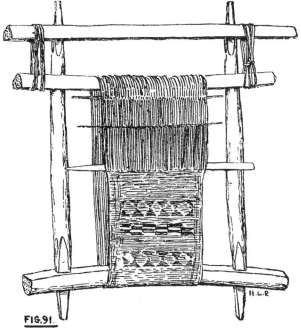

FIG. 91.
TYPE OF VERTICAL COTTON LOOM , WEST AFRICA

wooden model of two women weaving, found in a tomb at Beni Hasan,[1] is unfortunately on too small a scale to be of any assistance.[2]

---

[1] *Burial Customs of Ancient Egypt*, London, 1907.
[2] All these Egyptian Looms are illustrated in *Ancient Egyptian and Greek Looms*, already quoted.

The Madagascar loom shows Oriental influence, but I think that when Grandidier says that the loom " of the Malagashes is identical with that of the Indo-Oceanic peoples "[1] he goes much too far. It seems certain, however, that it has crossed from the island to the mainland of Africa, and in extending northward met another, a similar loom, coming south from Egypt or Somaliland—its extension westward along the Mediterranean and the Sahara being no doubt due to Arabic-Berber migrations.

FIG.91A. WOMEN WEAVING A JERRI ' FROM JEAN POMMEROL'S AMONG THE WOMEN OF THE SAHARA. LONDON 1900 P.299.

3. *The Vertical Cotton Loom.*—We now come to the vertical cotton loom on which plain and pattern cloths are woven. The illustration, Fig. 91, gives its chief characteristics, as it can be seen at the present day, on the West Coast in Abeokuta, Opobo, etc. In Figs. 91A and 91B we have it as met with at the present, in perhaps a more original form, in Algeria. It is everywhere worked by women only. Miss Gehrts[2] mentions that at Bafilo, the only place she seems to have observed the loom, the women weavers had a guild such as the men weavers have elsewhere.

[1] *Ethnographie de Madagascar*, Paris, 1908, p. 63, footnote.
[2] *A Camera Actress in the Wilds of Togoland*, London, 1915, pp. 93-4.

The West Coast modification consists of a square frame made up of an upper and lower piece of palm leaf mid-rib or stem into which are fixed two uprights ; occasionally, instead of the ends of the uprights passing through holes in the upper and lower ribs, they are merely lashed on to the latter, Fig. 92. The

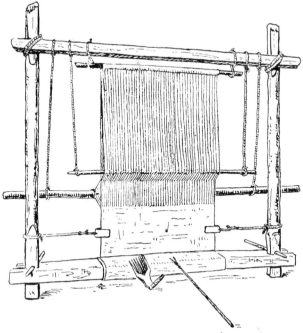

FIG. 91B. VERTICAL WOMENS HANDGRIP LOOM. FROM FRANZ STUHLMANN'S AUSFLUG IN DEN AURES, HAMBURG. 1912. AFTER B FATAH'S METHODE DIRECTE POUR L'ENSEIGNMENT DE L'ARABE PARLE. ALGER. 1904.

lower rib forms the breast beam, which is sometimes furnished with a supplementary rod A, Fig. 93 ; the upper rib itself occasionally forms the warp beam, but usually another rib suspended below it does this. As the palm leaf mid-rib employed does not possess much rigidity, the two beams sag towards each other when the warp is *beamed* (*i.e.*, put on to the loom). On the West Coast the warp is continuous, but in Algeria this is not the case. The weaving proceeds from below upwards. Generally the heddle consists of two very thin pieces of cane with spiral leashes intertwined, Fig. 94.

The Bankfield specimen of this class of loom is from Abeokuta, having been obtained there in 1904 by Mr. Cyril Punch, the donor. It is without the upper frame bar. Its dimensions are : length, beam to beam inclusive, 53 inches (or 1·35 m.) ; width of web, 27 inches (or 69 cm.) ; 26 picks to the inch (or 10 to the cm.) ; and 84 warps to the inch (or 33 to the cm.). The shed stick, like the two beams, is of palm leaf mid-rib. There are two rods which might be taken for ordinary laze-rods, and while they do to a certain extent function as such, the laying-out of the warp indicates them as pattern rods and similar to the

49

pattern threads in the Peruvian loom, Fig. 40, although in the web so far as it is woven there is no weft pattern. These rods are 1 by ½ inch (or 2·5 by 1·2 cm.), in thickness. The heddle, Fig. 95, contrary to the usual, consists of one strip of cane, ¼ inch (or 6 mm.), thick, and a piece of blue cord. It is raised by hand. When the pick is made and the heddle is dropped, the tension of the warp should be sufficient to bring down the warp threads 1, 3, 5, 7, etc., into the same plane as the warp threads 2, 4, 6, 8, etc., but owing to the want of rigidity in the beams already referred to, this only takes place in a very modified manner, and a raiser or picker-up, Fig. 96, for the warps, 2, 4, 6, 8 has to be brought into use. It consists of a thin rod of wrought iron (not wire) hafted into a suitably shaped piece of wood. The spool, 31½ inches (or 80 cm.), long, belongs to type A*a*1.

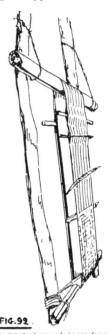

FIG. 91C TAPESTRY FRAME FROM MS. CODEX BY RABANUS MAURUS. IX CENT. (AFTER A. SCOLE, ENCYCLOP BRIT 19TH ED)

FIG. 92.
ILORIN WOMAN'S LOOM (APROX-IMATE REPRODUCTION) FRO-BINIUS' VOICE OF AFRICA. SIDE VIEW

The weft is generally stouter than the warp, which is laid alternately in varying breadths of brown and white which gives the striped pattern. The temple consists of two flat thin pieces of cane rind ⅝ inch (or 1·6 cm.), wide, Fig. 97, both ends of both pieces tapering to a point.

In the looms of this class from Opobo (Manchester, Salford, Liverpool and Glasgow Museums), from Southern Nigeria (Imperial Institute) and from Akweta, Lower Niger (Liverpool Musuem), there is brought into use a rod which is carved "herring bone" fashion on the surface, Fig. 98. Its function appears to be that of a laze rod and not that of a pattern rod : it would be of use mostly in long and broad weaving to prevent warp entanglement.

Coloured geometrical weft pattern weaving on these looms has reached a high pitch of excellence, all things considered ; blue, yellow, red, and white yarn being used over a blue warp and weft with generally a few inches of coloured warp at both selvedges. The pattern is woven on top of the plain web as the latter

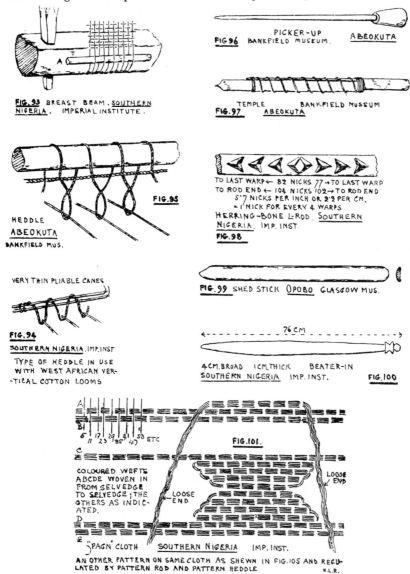

FIG.93 BREAST BEAM. SOUTHERN NIGERIA. IMPERIAL INSTITUTE.

PICKER-UP
FIG.96 BANKFIELD MUSEUM. ABEOKUTA

TEMPLE BANKFIELD MUSEUM
FIG.97 ABEOKUTA

HEDDLE
ABEOKUTA
BANKFIELD MUS.

FIG.95

TO LAST WARP← 82 NICKS 77→TO LAST WARP
TO ROD END ← 104 NICKS 102→TO ROD END
5·7 NICKS PER INCH OR 2·2 PER CM.
= 1 NICK FOR EVERY 4 WARPS
HERRING-BONE L-ROD. SOUTHERN NIGERIA. IMP. INST.
FIG.98

VERY THIN PLIABLE CANES

FIG.94
SOUTHERN NIGERIA. IMP.INST

TYPE OF HEDDLE IN USE WITH WEST AFRICAN VER- -TICAL COTTON LOOMS

FIG.99 SHED STICK OPOBO GLASGOW MUS.

76 CM

4CM.BROAD 1CM.THICK BEATER-IN
SOUTHERN NIGERIA IMP.INST. FIG.100

FIG.101.

COLOURED WEFTS ABCDE WOVEN IN FROM SELVEDGE TO SELVEDGE; THE OTHERS AS INDIC- ATED.

LOOSE END

LOOSE END

"PAGN" CLOTH SOUTHERN NIGERIA IMP. INST.

ANOTHER PATTERN ON SAME CLOTH AS SHEWN IN FIG.105 AND REGU- LATED BY PATTERN ROD AND PATTERN HEDDLE. H.L.R.

proceeds, and is woven right across the web or in part only as required ; if in part only the ends of the weft hang down as shown in Fig. 101 until further required. For this sort of pattern weaving the worker is guided by the special way in which the warp is laid out. Every third, fourth, sixth or twelfth warp, as the case may be, is made to pass over the pattern rod as shown in Figs. 102, 103 and 105, and

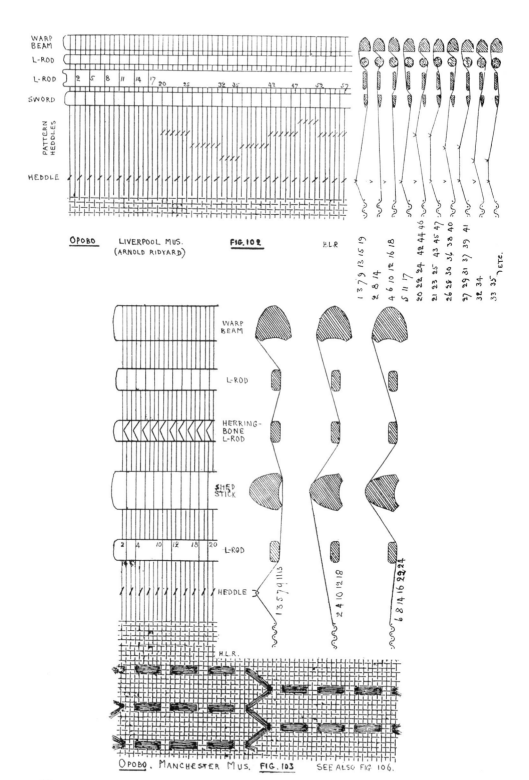

WARP BEAM
L-ROD
L-ROD   2   5   8   11   14   17   20   25   32   35   43   47   52   57
SWORD
PATTERN HEDDLES
HEDDLE

OPOBO   LIVERPOOL MUS.   FIG. 102   H.L.R
(ARNOLD RIDYARD)

1 3 7 9 13 15 19
2 8 14
4 6 10 12 16 18
5 11 17
20 22 24 42 44 46
21 23 25 43 45 47
26 28 30 36 38 40
27 29 31 37 39 41
32 34
33 35) ETC.

WARP BEAM
L-ROD
HERRING-BONE L-ROD
SHED STICK
L-ROD   2   4   10   12   18   20
HEDDLE

1 3 5 7 9 11 13
2 4 10 12 18
6 8 14 16 22 24

H.L.R.

OPOBO. MANCHESTER MUS. FIG. 103   SEE ALSO FIG 106.

52

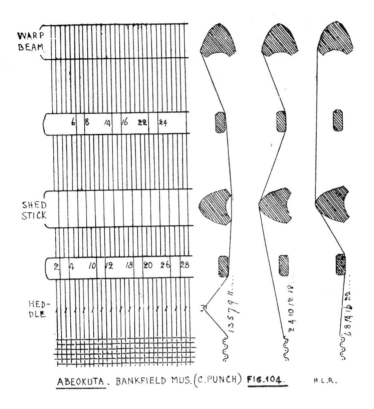

WARP BEAM

6 8 14 16 22 24

SHED STICK

2 4 10 12 18 20 26 28

HEDDLE

1 3 5 7 9 11

2 4 10 12 18

1 8 14 16 22

ABEOKUTA. BANKFIELD MUS. (C.PUNCH) FIG.104. H.L.R.

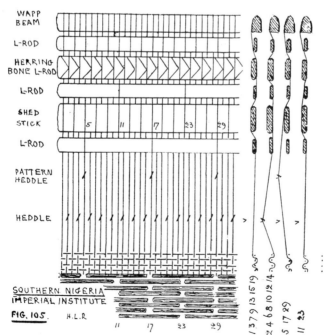

WARP BEAM

L-ROD

HERRING BONE L-ROD

L-ROD

SHED STICK

5 11 17 23 29

L-ROD

PATTERN HEDDLE

HEDDLE

SOUTHERN NIGERIA
IMPERIAL INSTITUTE
FIG. 105. H.L.R

11 17 23 29

1 3 7 9 13 15 19

2 4 6 8 10 12 14

5 17 29

11 23

53

in order to ascertain at which warp the pattern weft is to be inserted or withdrawn the weaver must apparently run his finger from that warp on the pattern rod down to the web, and where that warp passes into the web he will insert or withdraw his spool as the case may be. One loom, Fig. 102, from Opobo (Liverpool Museum) is provided with four pattern heddles as well as a pattern rod.

According to van Gennep[1] the designs are taken from domestic objects, and he mentions particularly that one is taken from the pulley-block of a treadle loom. On the other hand Pommerol,[2] who no doubt had better access to the womenfolk than he should have had, speaking of an excellent old woman, says : " El Haj teaches novices the art of casting the threads of the weft [*sic.*, should be warp.—H. L. R.] from one peg to another and arranging these threads vertically in the primitive looms, made of wood, string and reeds. She teaches them how to dye wool and how to mix the different shades of colour ; but one thing she jealously guards, and that is the secret of the hieroglyphics, those mysterious and cabalistic designs, such as

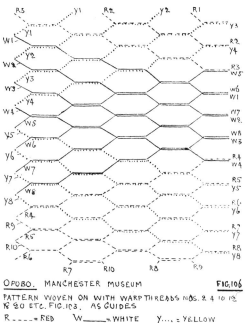

OPOBO. MANCHESTER MUSEUM   FIG.106

PATTERN WOVEN ON WITH WARP THREADS NOS. 2 4 10 12 & 20 ETC. FIG.103. AS GUIDES

R - - - - = RED   W _____ = WHITE   Y... = YELLOW

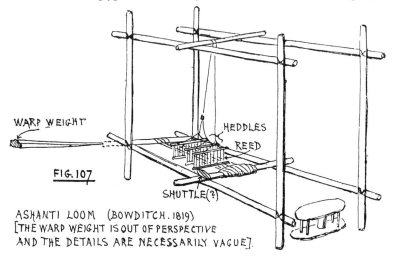

WARP WEIGHT   HEDDLES   REED

FIG.107

SHUTTLE(?)

ASHANTI LOOM (BOWDITCH.1819) [THE WARP WEIGHT IS OUT OF PERSPECTIVE AND THE DETAILS ARE NECESSARILY VAGUE].

squares, zig-zags and arabesques, which represent sometimes an object, sometimes an idea and sometimes a phrase. Only to a few initiated does El Haja teach, and

[1] *Etudes d'Ethnographie Algerienne*, Paris, 1911, p. 100.   [2] *Op. cit., p.* 298.

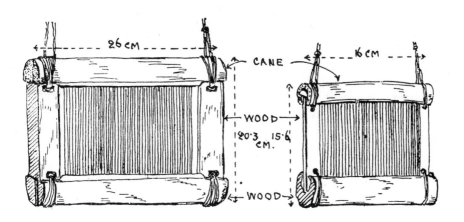

CANE

26 CM

16 CM

WOOD

18.3 | 15.6 CM.

WOOD

REED AND SHUTTLE

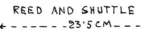

←----- 23.5 CM -----→

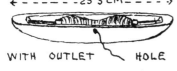

WITH OUTLET HOLE

ABEOKUTA (C.PUNCH) 1903

FIG. 108

REED AND SHUTTLE

←----- 22.8 CM -----→

NOT PROVIDED WITH OUTLET HOLE.

KWITTA. (ARN. RIDYARD) 1915

BANKFIELD MUSEUM.

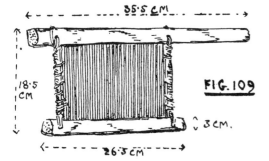

35.5 CM

18.5 CM

3 CM.

26.5 CM

FIG. 109

A

A

REED OF TRIPOD LOOM WITH EXTENSION TO FORM A HANDLE. WITHOUT ANY ROD TO JOIN THE TWO HORI-ZONTALS (UPPER & LOWER PORTIONS).

TIKONKO VILLAGE (MENDE PEOPLE), SIERRA LEONE

(REV. W.T. BALMER) 1904.
BANKFIELD. MUS.

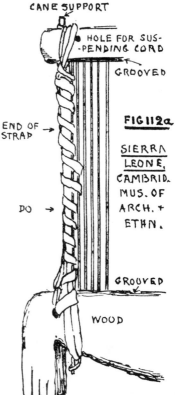

CANE SUPPORT

HOLE FOR SUS--PENDING CORD

GROOVED

END OF STRAP

FIG 112a

SIERRA LEONE, CAMBRID. MUS. OF ARCH. + ETHN.

DO →

GROOVED

WOOD

H.L.R.

55

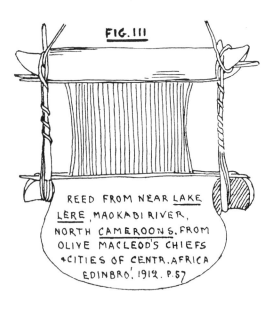

FIG. III

REED FROM NEAR LAKE
LERE, MAOKABI RIVER,
NORTH CAMEROONS. FROM
OLIVE MACLEOD'S CHIEFS
+ CITIES OF CENTR. AFRICA
EDINBRO', 1912. P.57

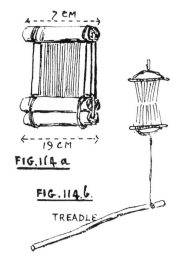

2 CM

19 CM

FIG. 114 a

FIG. 114. b

TREADLE

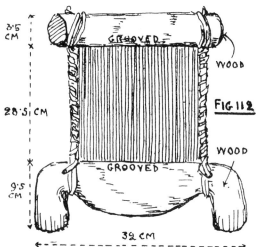

3·5 CM

GROOVED

WOOD

28·5 CM

FIG 112

9·5 CM

GROOVED

WOOD

32 CM

REED. SIERRA LEONE
CAMBRIDGE MUS. ARCHEOL. & ETHNOL.

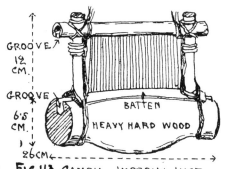

GROOVE
12 CM.

GROOVE
6·5 CM.

26 CM

BATTEN

HEAVY HARD WOOD

56 FIG. 113 GAMBIA. IMPERIAL INST.   H.L.R.

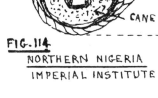

CANE

ENDS OF
BATTENS
HOLDING
THE REEDS
IN POSITION

END OR
SIDE PIECE

21
CM

WOOD

CANE

FIG. 114
NORTHERN NIGERIA
IMPERIAL INSTITUTE

that grudgingly, this ancient writing, which she herself does not understand enshrouding as it does the thoughts of races long since passed away."

The loom is used for providing a variety of articles, and is also used as a tapestry loom and so carrying us back to the Middle Ages.   A. S. Cole illustrates what appears to me to be the same loom in an article on Tapestry,[1] Fig. 91C, from a IXth Century MS.   As a tapestry loom it is the same apparatus used for making rugs in India (there is a specimen in Bankfield Museum), and as an ordinary cloth loom it is presumably the same as the Ancient Egyptian loom depicted in the tombs of Thot-nefer and Nefer-ronpet, already referred to, although in the latter one of the looms appears to be served by a woman.

Stuhlmann,[2] following Frobenius, proposes to call this form of loom the grip loom because the shed is made by gripping the heddle instead of the shed being made by means of treadles.   The term is suitable if confined to looms in which there is really a gripping of the heddles, but cannot be applied to all non-treadle looms, for the latter include the Fixed Heddle Looms.

4.   *The Horizontal Narrow Band Treadle Loom.*—The narrow band treadle loom, which is a fully developed loom supplied with reed, harness, treadles, etc., may be seen in two sub-forms as it were : (*a*) the one furnished with a rectangular frame on the plan of our hand looms, and (*b*) the other furnished with a *tripod* frame.

The first writer to give us an illustration of the rectangular frame loom was T. E. Bowditch,[3] who tells us : " The Ashantee loom is precisely on the same principle as the English ; it is worked by strings held between the toes ; the web is never more than four inches (10 cm.) broad."   His illustration is reproduced here, Fig. 107, *minus* the weaver, who does not add to the clearness of it.   The illustration depicts a rectangular frame consisting of four uprights, three top beams, and four lower beams ; the warp extends beyond the frame and appears to be made taut to a heavy stone, acting as a sort of anchor while the cloth is rolled up on a breast beam ; there are two heddles which look very much like the reed (beater-in), and in spite of his remark he does not show the string connecting the lower shafts of the heddles with the weaver's toes ; it is possible that there is an indication of a shuttle.   The narrow pieces of cloth (band) produced on this loom vary from 2 to 6 inches (5 to 15 cm.), in breadth.   The reed, Fig. 108, is the best part of it, and generally consists of a more or less square hardwood frame 6 inches by 7 inches (15 by 18 cm.), or thereabouts ; all four sides are equally thick, and when fitted together and tied up the whole is fairly rigid.   The reed is also frequently made with the side pieces much lighter than the top and bottom, so that it would be too light to swing back into position after beating-in, to obviate which the bottom wood is made heavier than the top one, or other pieces of wood are fastened to it, as shown in Figs. 111, 112, 113, and 114.   The reed is suspended by cord and not by side battens like our sley side supports.   The heddles are likewise suspended by a cord which passes over an ill-formed pulley, Fig. 119, the roller of which is

[1] *Encyclopaedia Britannica*, xiith ed.   [2] *Aures*, p. 117.
[3] *Mission from Cape Coast Castle to Ashantee.*   4to.   London, 1819.

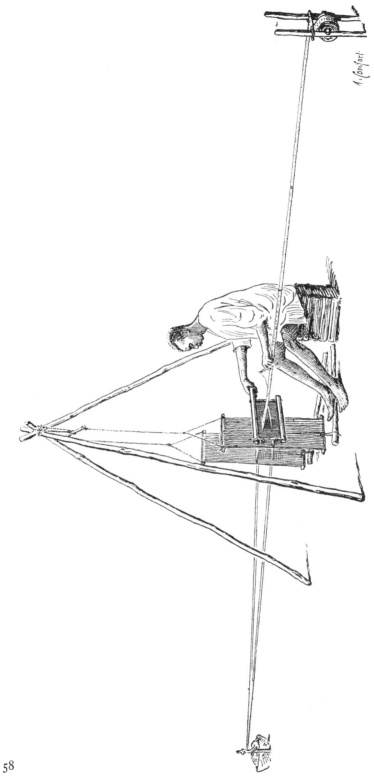

A. Comfort.

FIG. 110. HORIZONTAL NARROW BAND LOOM.—TRIPOD FORM. FROM A SPECIMEN IN BANKFIELD MUSEUM FROM TIKONKO, MENDE, SIERRA LEONE (REV. W. T. BALMER, 1904) & A PHOTOGRAPH.

frequently an old sewing yarn reel (or bobbin). The leashes of the heddles are mostly of twisted cotton. A shuttle is in use, and frequently it is *not* provided with a weft paying-out hole. The warp beam's place is taken by a heavy stone or anchor, while the breast beam consists of a thin cylindrical stick on which the cloth is wound by means of a wooden pin passed through the end, the point of which presses against the seat of the weaver to prevent unwinding. The heddles are drawn downwards by means of cords which end in a wood or bone disc or a short transverse piece of wood, which is grasped between the big and second toes of

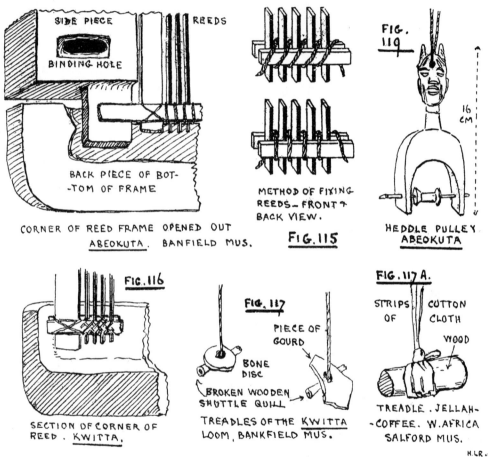

SIDE PIECE   REEDS

BINDING HOLE

BACK PIECE OF BOT-
-TOM OF FRAME

CORNER OF REED FRAME OPENED OUT
ABEOKUTA. BANFIELD MUS.

METHOD OF FIXING
REEDS — FRONT &
BACK VIEW.

FIG. 115

FIG. 119

16
CM

HEDDLE PULLEY
ABEOKUTA

FIG. 116

SECTION OF CORNER OF
REED. KWITTA.

FIG. 117

PIECE OF
GOURD

BONE
DISC

BROKEN WOODEN
SHUTTLE QUILL

TREADLES OF THE KWITTA
LOOM, BANKFIELD MUS.

FIG. 117 A.

STRIPS   COTTON
OF   CLOTH

WOOD

TREADLE. JELLAH-
-COFFEE. W. AFRICA
SALFORD MUS.

H.LR.

the weaver, Figs. 117, 117A, or in some cases the cord ends in a loop into which the weaver inserts his big toe. The weaver sits facing the breast beam.

The student naturally compares this loom with our own hand looms, and it has some resemblance to the broad, well-developed hand loom which is found the whole length of North Africa, inclusive of Eygypt. But if the similarity be there, then also we must admit that this African loom is a very degenerate representative, judging by the complete specimens in Bankfield Museum, in the Liverpool Museum, and by the portions to be found in our museums elsewhere. The frames are extremely flimsy, ill-fitted together, and slovenly rather than crude in their details. The frequent omission of a pay-out hole in the shuttle is very probably a sure sign of decay. On the other hand, it is doubtful if the connection making the corner of the better sort of reed in Fig. 115 can be considered anything but of very recent origin, and not by any means African.

Perhaps still more degenerate is the tripod form, Fig. 110. In this the reed and treadles are suspended from the jointing of the three poles which make up the tripod. It is further characterized by an extension of the upper portion of the reed into a handle, Fig. 109, which is grasped in the right hand of the weaver, who sits on the right-hand side of the web, and not with his work straight in front of him. Why the weaver should take up such a position is not obvious, for one would

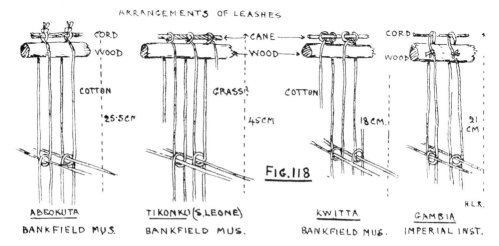

ARRANGEMENTS OF LEASHES

FIG. 118

| ABEOKUTA | TIKONKO (S. LEONE) | KWITTA | GAMBIA |
| BANKFIELD MUS. | BANKFIELD MUS. | BANKFIELD MUS. | IMPERIAL INST. |

think that the beatings-in by the reed would have a tendency to get away from the right angle to the warp, but I am unable to trace any such irregularity in the fabrics examined. Owing to the absence of wooden side pieces, the reed is anything but rigid. The heddles are supported from a whipple tree ; the leashes are of fine twisted grass or similar filament. The cords drawing down the heddles are attached to one end of each of two sticks (the treadles), the other two ends touching the ground, which gives the sticks an oblique position. Judging by illustrations of negroes at work, the weaver does not keep his feet one on each treadle, but uses one foot alternately for both treadles. The place of the warp beam is taken by a post fixed vertically in the ground, and the yet-to-be-used warp is rolled partly round it and placed in a basket at the side. The web is wound round a horizontal stick (what would otherwise be the breast beam) placed against a pair of uprights ; but I am not conversant with the details. A spool of the A1 type is used and *not* a shuttle, and, like the rectangular loom, it is tended by men weavers only.

In connection with this loom we have an interesting specimen of the Warp-Laying Frame in the Horniman Museum, Fig. 120. It consists of a frame of which the two uprights of roughly squared soft wood are connected by two transverse flat pieces of wood, each of which is provided with a longitudinal slot. As near as possible to the middle of the frame a thin rod of wood, A, is inserted through the uprights parallel with the flats. A strip of leaf is then taken, bent at its middle over the bar, and its ends woven through the slot in the lower flat and knotted together below. A similar strip of leaf is passed through the loop of the first strip bent under the bar, its ends passed upwards, woven through the slot of the upper flat and knotted together above. This work is repeated until the frame is full of these leashes—in this frame there are 41 double leashes—whereupon the bar is removed and a separate warp thread laid through every space, between the bends in the leashes, left vacant by the bar. The two uprights are then knocked off and the two slotted flats and leashes and warps transferred to the heddle frame ready for

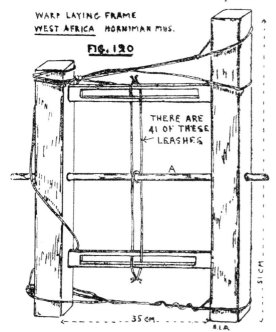

WARP LAYING FRAME
WEST AFRICA HORNIMAN MUS.

FIG. 120

THERE ARE 41 OF THESE LEASHES

A

35 CM.

weaving. Length from top to bottom over all of slotted bars 12 inches (30·5 cm.), width between two posts 10 inches (12·5 cm.).

J. Büttikofer informs us[1] that as the weaver " proceeds with his work he pushes the *whole* of the apparatus forwards to the right, the warp not being moved at all." The immovable warp is a characteristic of some of the fixed heddle looms (see Figs. 89 and 89A), but whether the women of Neh, in the extreme eastern part of Persia near Afghanistan, move forward their tripod frame, the warp remaining in position, or whether Singalese weavers, who also use a tripod frame, do so or not, I have not been able to ascertain, but I think the tripod must be moved forward in all cases where the warp and breast beams are pegged to the ground, and the warp is not continuous.

The tripod is found in other parts of Africa besides the West. Miss A. Breton writes me : " At Luxor, in the market, I saw [in 1909] a man weaving with a most primitive gipsy kettle contrivance—three legs like a gipsy kettle—the result was as good as could be wished." In the course of ages a similar tripod loom may have travelled across the Continent, meeting in the more western portion of West Africa with the loom which Dr. Harrison has suggested that the Portuguese possibly introduced in the sixteenth century[2]; the two gradually merged into each other, giving us the oblong frame with a fair reed, Fig. 108, in the more western portion

[1] *Reisebilder aus Liberia*, Leiden, 1890, ii, p. 283.
[2] *Horniman Museum Handbook*, " Domestic Arts," Part II, p. 50.

and keeping the tripod with a poorer reed, Fig. 109, in the more eastern portion, but in both cases adopting the harness for the heddles and treadles. This would account for the noticeable degenerate appearance of the looms. This view also coincides with that held by Sir H. H. Johnston,[1] in so far that the loom is not indigenous here, but was introduced from the East, and was primarily due to the advance of the Arabs in the ninth century A.D.

5. *The Pit Treadle Loom.*—This is the common Hindu loom at which the weaver sits on the edge of a specially constructed hole in which his feet work the treadles, and is rather a method of working a loom than a distinct form of loom. It is met with largely in the green mountains district of Oman, Arabia, half-way between India and Africa. Colonel S. B. Miles in describing it,[2] says : " The weaver sits and works at it in a shallow pit, with half his body below the surface." In Africa it is found among the Gallas[3] and contingent peoples. In Bankfield Museum we have a fringe-making loom, the gift of Mr. Wm. Myers, lecturer in the Manchester Municipal Technical College, which was obtained near Khartum by Mr. C. S. Rhodes, and in forwarding the specimen, Mr. Rhodes sent a sketch showing the native weaver seated with his feet in a hole working the treadles.

6. *The Mediterranean Loom.*—I call this form the Mediterranean loom for want of a better name. It is the usual rectangular heavy framed loom with roller warp and breast beams, very similar to that of our few remaining hand weavers' looms.

It appears to be distributed along the north coast of Africa from Algiers to Egypt and, perhaps, somewhat up the Nile Valley, although the illustration in Dr. John Garstang's work[4] seems to depict a loom of a different type. Neither, however, has the warp bunched together over the head of the weaver and weighted behind him, as is the case in Syria.[5]

Like the pit treadle loom, it cannot in any way be considered African.

7. *" Carton " Weaving.*—This method of obtaining bands, girdles, sashes, etc.—*i.e.*, more or less narrow fabrics—can, strictly speaking, hardly be called weaving. Its former use in Egypt has been so well described and illustrated in a sumptuous work[6] by Messrs. A. van Gennep and G. Jéquier that it need only be referred to here. In his *Etudes* already referred to (see p. 54) van Gennep gives the distribution of this tool as extending along the north coast of Africa from Tangiers to Tunis inclusive, and on the banks of the Lower Nile.

The Map.—In the accompanying map, Fig. 121, I have endeavoured to convey to the student some idea as to the distribution of the various forms of looms found

---

[1] *Liberia*, London, 1906, ii, pp. 1016-18.
[2] *Geog. Journ.*, November, 1900, xviii, p. 475.
[3] F. J. Bieber, *Globus*, March 8th, 1908.
[4] *The Burial Customs of Ancient Egypt*, London, 1907, Fig. 133, p. 134.
[5] I gather this characteristic from illustrations of Syrian looms, kindly sent me by Dr. Harvey Porter, of the American Baptist Mission, Beirut.
[6] *Le Tissage aux Cartons et son utilisation decorative dans l'Egypte Ancienne*, Neuchatel, Suisse, 1916.

in Africa at the present day. The attempt must be regarded as strictly tentative only, for, while the main positions are, I think, fairly correctly placed, details of the extension of each individual form are still lacking.

With the exception of the vertical mat loom, which may possibly be indigenous to the heart of Africa, but about which we have not sufficient evidence to decide at present, and of the vertical cotton loom, which may have had its birth

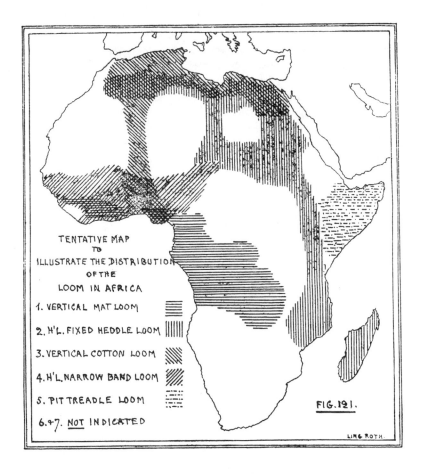

TENTATIVE MAP
TO
ILLUSTRATE THE DISTRIBUTION
OF THE
LOOM IN AFRICA

1. VERTICAL MAT LOOM
2. H'L. FIXED HEDDLE LOOM
3. VERTICAL COTTON LOOM
4. H'L. NARROW BAND LOOM
5. PIT TREADLE LOOM
6.+7. NOT INDICATED

FIG. 121.

LING ROTH.

in Egypt, all the other five forms are introduced. The fixed heddle loom appears to have entered Africa both in the north-east *via* Arabia and in the south-east *via* Madagascar. The horizontal narrow band treadle loom came possibly from Portugal, and the pit treadle loom was probably imported from India *via* Arabia. The Mediterranean or Asiatic treadle loom and the " carton " loom probably found their way in *via* the Mediterranean, if the latter was not indigenous to Egypt.

6. INDONESIAN LOOMS.

THE Indonesian loom belongs to the Pacific type of loom, two forms of which, the Ainu and American, have already been described in Sections 3 and 4. There appear to be three forms of loom in Indonesia, taking the area in a wide sense. They are the Dusun and Iban (Sea Dyak) loom, the Ilanun and Igorot transition loom, and the Cambodia and Malay loom. They all merge more or less into one another, and are therefore to be taken rather as various stages in the development of the loom than as perfectly distinct forms.

The following table gives dimensions and capacity of five such looms examined by me :—

Name of Tribe from whom obtained.	Museum where now placed.	Length, Beam to Beam, inclusive.		Width of Web.		No. of Warp per—		No. of Picks per—		Back Strap.	Material.	Heddle Leashes, continuous.	Repeat of Lay of Warp.	Form of Weft Carrier.
		In.	Cm.	In.	Cm.	In.	Cm.	In.	Cm.					
Dusun	British...	26½	67	12	30·5	38	15	28	11	Coarse cloth	Cotton	Spiral ...	Twos ...	Ab 1.
Iban ...	,, ...	10 9/16	27	11⅛	28	110	43.4	37	14.7	Raw hide	,,	,, ...	,, ...	Missing
,, ...	Horniman	23½	60	9 11/16	24·5	—	—	17	6·6	Rotan mat	,,	,, ...	Sixes ...	Ab 1.
,, ...	Liverpool	30	76	6½	16·5	118	74	40	16	Missing	Silk ...	Alternate over-lapping	Eights ...	Missing
,, ...	Royal Scottish	22½	57	5¼	13.3	—	—	—	—	,,	,,	,,	Twenties	Ba.

The most primitive of these is the Dusun and Iban loom (Figs. 122 and 123). It consists of a warp beam attached to two upright posts, a breast beam attached to a back strap, several laze rods, a shed stick, one " single " heddle, a beater-in, a temple, and a spool. The warp is continuous, and the weaver sits on the floor. The breast beam is almost in the weaver's lap, whence the warp rises at an angle of about 350° up towards the warp beam (Fig. 123). As there is only a " single " heddle there are no treadles nor does there appear to be any special loom frame, and the loom can be set up wherever there are a couple of suitable posts and a

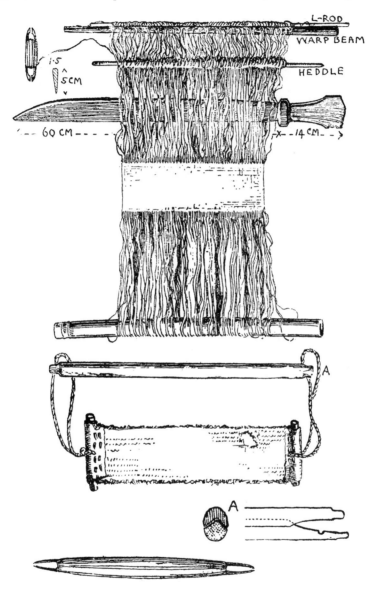

FIG. 122. DUSUN LOOM, BORNEO. BRIT. MUS.

suitable floor or platform. According to Hose and McDougall,[1] " The weaving is done only by the women, though the men make the machinery employed by them."

[1] *The Pagan Tribes of Borneo*, Lond., 1912, I, p. 221.

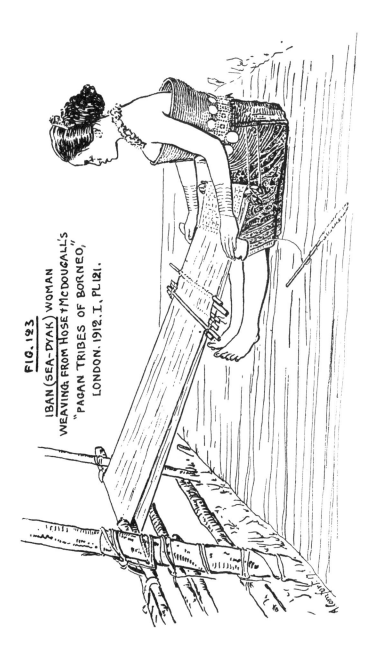

FIG. 123

IBAN (SEA-DYAK) WOMAN
WEAVING. FROM HOSE & McDOUGALL'S
"PAGAN TRIBES OF BORNEO",
LONDON. 1912. I, PL 121.

Most of the webs commence with two heading rods. The Horniman Museum specimen has an extra rod over and parallel with the heddle rod (Fig. 124), evidently to be used as a handle or raiser. In most of the looms the warp and weft are both double (" sisters "). The British Museum specimen is provided with a temple[1]

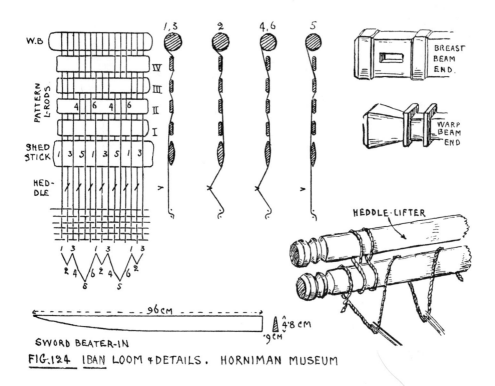

FIG. 124 IBAN LOOM & DETAILS. HORNIMAN MUSEUM

and has two warp beams ; it has an insignificant brocade pattern woven-in on the wrong side, as well as a warp pattern scheme extending the whole width of the cloth, thus :—

Selvedge mm.	Red ... 57	L. blue 4	Yellow 4·5	Gr. yel. 5	Dk. blue ... 2·5
Dark blue ... 2·5	Gr. yell. 3·5	Yellow 4·5	L. blue 6	Red ...59	Selvedge
Light blue ... 2·5	Yellow 4·5	Gr. yell. 3·5	Dk. blue 14	Gr. yel. 3·5	Total, 277 mm.
Yellow ... 2·5	L. blue 4	Red ... 58	L. blue 6	Yellow 2·5	(= 105 ins.)
Green-yellow 3·5	Dk. blue 14	Gr. yell. 3·5	Yellow... 4	L. blue 2·5	

The coloured warp is a characteristic of these looms. In the specimen in the Liverpool Museum the figured pattern is woven-in similarly on the wrong side and follows the laying of the warp, which repeats in eights as shown in Fig. 125, nearest to where the work has been left unfinished. There is another figure pattern further away (not shown), which does not agree with this warp-laying. The warp at the

[1] A loom from Sermata Island, between Timor and Timor-laut, in the British Muesum, is provided with a similar temple but quite flat, 12 mm. broad and 4 mm. depth of point.

selvedge begins with red and follows on with blue and white, then red in the centre for a width of 5¾ inches (14·5 cm.), and then white, blue, yellow and red to the opposite selvedge. The weft is gold, blue, white and green yarn.

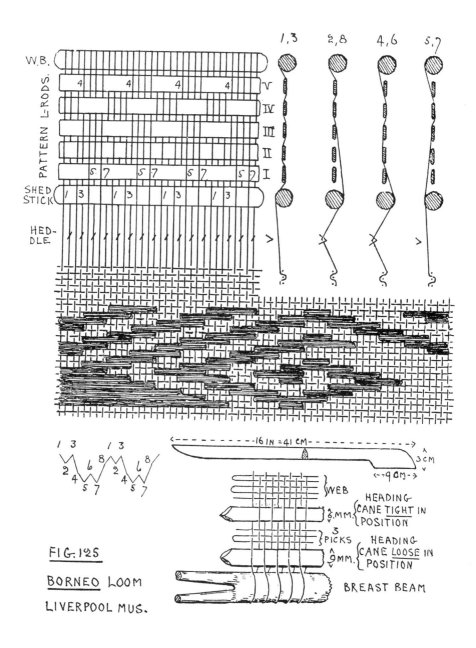

FIG. 125

BORNEO LOOM

LIVERPOOL MUS.

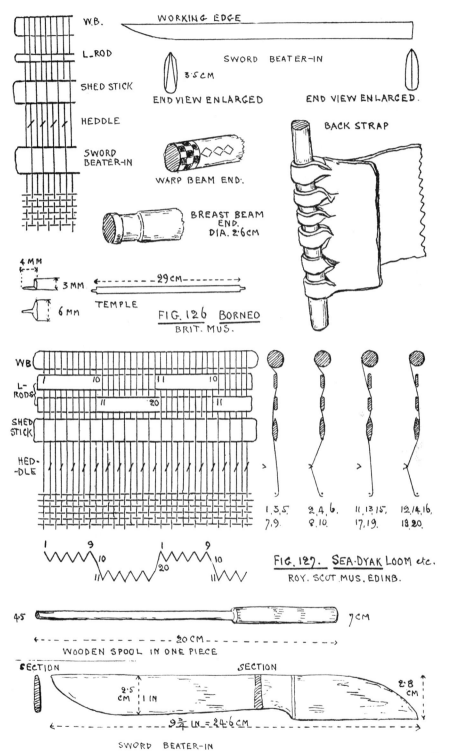

W.B.

L-ROD

SHED STICK

HEDDLE

SWORD
BEATER-IN

WORKING EDGE

SWORD BEATER-IN

3·5CM

END VIEW ENLARGED

END VIEW ENLARGED.

BACK STRAP

WARP BEAM END.

BREAST BEAM
END.
DIA. 2·6CM

4 MM

3 MM

6 MM

←- - - -29CM- - - -→

TEMPLE

FIG. 126 BORNEO
BRIT. MUS.

WB

L-RODS

SHED
STICK

HED-
-DLE

10   11   10

11   20   11

1.3.5.   2.4.6.   11.13.15.   12.14.16.
7.9.   8.10.   17.19.   18.20.

1   9
W\W\\10
1   9
W\W\\10
11   20   11

FIG. 127. SEA-DYAK LOOM etc.
ROY. SCOT. MUS. EDINB.

45   7CM

←- - - - - - - -20CM- - - - - - - -→
WOODEN SPOOL IN ONE PIECE

SECTION   SECTION

2·5
CM   1 IN   2·8
CM

←- - - - -9¾ IN = 24·6CM- - - - -→

SWORD BEATER-IN

69

A similar loom (Fig. 125A) is found in Sumatra, whence H. O. Forbes brought one from Moeara Doea in 1873, now in the British Museum. The particulars are :

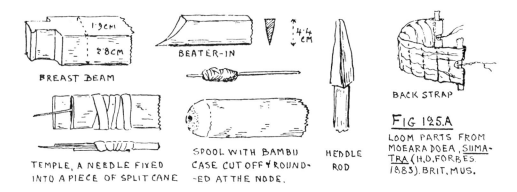

BREAST BEAM

BEATER-IN

BACK STRAP

TEMPLE, A NEEDLE FIXED INTO A PIECE OF SPLIT CANE

SPOOL WITH BAMBU CASE CUT OFF YROUND-ED AT THE NODE.

HEDDLE ROD

FIG 125.A

LOOM PARTS FROM MOEARA DOEA, SUMA-TRA (H.O.FORBES. 1883).BRIT.MUS.

length, beam to beam inclusive, $32\frac{1}{2}$ inches (or 82·5 cm.); width of web 17 inches (or 43 cm.); 136 warp to the inch (or 53·5 to the cm.); 32 picks to the inch (or 12·6 to the cm.). Warp and weft twisted ; leashes continuous, alternate, overlapping. Breast beam rectangular in section ; warp beam of wood $1\frac{3}{4}$ inches (or 4·5 cm.) in diameter. There is no reed. The temple consists of a flat piece of cane with needle inserted in a split at both ends, and reminds one of the similar American tool (Figs. 36 and 37). The warp is coloured in bands of red, yellow, and blue, a further pattern on blue ground being made in the centre, $4\frac{3}{4}$ by 5 inches (or 12 by 12.7 cm.), by means of white, red, crimson and yellow yarns, the ends of which are cut off on the surface when done with. There are two heddles, one for the general weave and one for making the border. The back strap is composed of a piece of bark, 15 by 5 inches (or 38 by 12.7 cm.), padded on the inside by cotton wool sewn into a bag of coarse cotton. The ends of the strap are strengthened by pieces of cane to which the beam ends are attached.

This form of loom is also found in the Philippine Islands, among the Ifugaos, Tingiaus, etc.,[1] in Assam, in parts of Burma among the Karens, and also in Tibet, but somewhat modified. Hose and McDougall[2] tell us as regards Borneo that weaving " is the only craft in which Ibans [Sea Dyaks] excel all other peoples," although my necessarily more limited experience leads me to the conclusion that Ilanun weaving far excels that of the Iban. Hose and McDougall continue : " Their methods [*i.e.*, those of the Iban] are similar to those of the Malay and have probably been learnt from them." Here, too, I must differ, for as we shall see directly, the loom used by the Malays proper is a more advanced article than that used by the Ibans, and if the Ibans had learnt from the Malays I think we are more likely to have found among them an imperfect or

[1] See Worcester, *Phillipine Jour. Sci.*, I, No. 8.
[2] *Op. cit.*, I. p. 220.

degenerate form of loom, rather than a more primitive one than that used by the Malays. In the same way that the house building of the Kenyahs, and the padi cultivation of the Klemantans, are both inferior to these arts as practised by the Kayans, from whom the Kenyahs and Klemantans are thought to have learnt the arts.[1] The Ibans probably brought their loom with them from Sumatra. The Dusuns are probably of Philippine origin, and hence the survival of this primitive loom amongst them. They possess a considerable amount of Chinese blood, and from this one would be inclined to think they might have adopted an improved method of weaving, in the same way as they have improved their cultivation by adopting the plough. But the Chinese who settled among them probably took Dusun wives to themselves, and as weaving is women's work in these parts, and there were no Chinese women to show them better, the primitive loom has survived ; and as a corollary, but as a side issue here, ploughing being men's work the Dusun were taught by the Chinese men how to plough, and that tool has been retained.

An observation of T. Chapman, quoted by me,[2] runs as follows : " At present there are only two kinds of looms :   the *tumpoh*, at which the weaver sits on the floor and uses his hands only ; and the *tenjak*, at which the weaver sits on a bench and uses hands and feet, the latter working the treadles. The cloths are much better and closer woven on the *tumpoh* looms. Both looms are picturesquely clumsy and the work slow." Here Chapman is referring to the Iban loom and to the Malay loom, which, as he indicates, show wide divergence from each other. The Iban may, no doubt, have learnt from the Malay in occasionally adopting the latter's loom, and to say that what they know of weaving they have learnt from the Malay can only refer to what they have learnt of weaving *on* Malay looms, while the superior work produced on their own looms shows they are not yet conversant with the methods of the later intruder.

The Bhotiyas loom, Darjeeling (Fig. 128), shows some advance on the Iban and Dusun loom in being provided with three single heddles instead of one. Apart from this and the heavy composite beater-in, it is similar to the Borneo specimens, but is also provided with a cloth beam, or second warp beam, according to one's point of view. The warp, which is spun wool, is only partly continuous and is arranged as follows : No. 1 warp starting from the cloth- or No. 2 warp-beam, goes its round over the breast beam across the three heddles, three shed sticks, laze-rod and warp beam until it reaches the second warp beam from the opposite direction, when it starts the return journey, getting back as No. 4 warp. The same laying holds good for No. 2 warp, which returns as No. 3. The length of the loom, *i.e.*, breast beam to first warp beam inclusive, is 9 feet 10 inches (or 3 m.), and the width of the cloth is 17 inches (or 43 cm.). The shed sticks are bevel-edged, about $1\frac{1}{2}$ inches (or 3·8 cm.) wide and 27 inches (or 69 cm.) long, the middle one

---

[1] Hose and McDougall, II, p. 244.
[2] *Natives of Sarawak*, II. p.30.

being slightly curved like a boomerang. The laze rod consists of a piece of cane about $\frac{1}{2}$ inch (or 12·7 mm.) in diameter, round which every warp is wound once

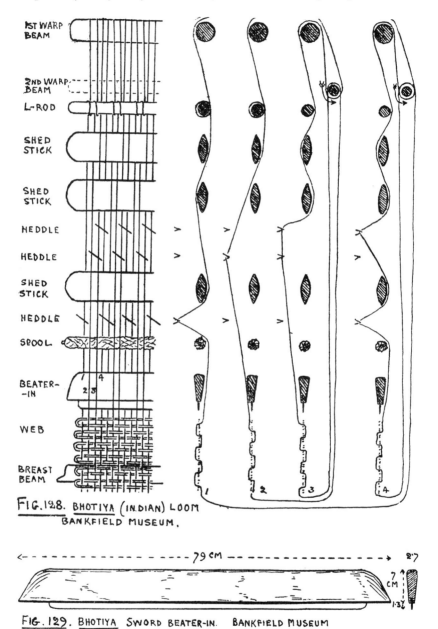

FIG. 128. BHOTIYA (INDIAN) LOOM BANKFIELD MUSEUM.

FIG. 129. BHOTIYA SWORD BEATER-IN. BANKFIELD MUSEUM

so that the rod can be rolled backwards and forwards, and still keep the threads in position. There are three spools of the Aa form about $\frac{3}{8}$ inch (or 9·5 mm.) in diam. and 24 inches (or 61 cm.) long, and when completely filled the ends are likewise covered.[1]   The weft is double ("sisters") and the warp is single.   The heddle leashes, which are spiral, naturally require to raise two warp threads in every leash to make the pattern.   The beater-in (Fig. 129) consists of a heavy piece of wood 31 inches (or 79 cm.) long by $3\frac{1}{4}$ inches (or 8·2 cm.) wide, very thick at the back and tapering to the front, where it is provided with a piece of wrought iron (*not* hoop iron), let in lengthwise and protruding about $\frac{1}{2}$ inch (or 12·7 mm.) beyond the wood.   This blade, like the back itself, is wedge-shaped in section.   The heaviness of this tool may be necessary as a very coarse wool has to be beaten in. The edge of the iron, the back of the beater-in, and both edges of all three shed sticks, are deeply serrated from friction in the working.   The temple is cut out of a thin strip of cane shouldered and pointed at both ends.

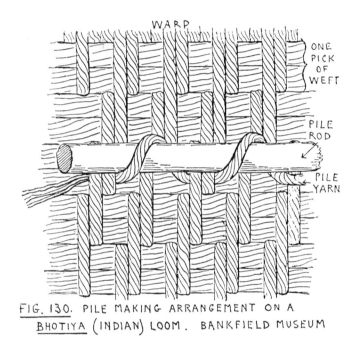

FIG. 130. PILE MAKING ARRANGEMENT ON A BHOTIYA (INDIAN) LOOM. BANKFIELD MUSEUM

Another Bhotiya loom, which I saw at work at the Coronation Exhibition in London in 1910, is now likewise in Bankfield Museum, and is fitted up for making rugs or pile cloth.   It is provided with a ball of weft instead of a spool of weft.   In other respects the two looms are similar.   The length from beam to beam inclusive was about 18 feet (about 5·5 m.), with continuous warp, and the angle of rise of the warp from the weaver was somewhat under 30°.   The method of inserting the pile is shown in Fig. 130.   It may be likened to that of a heddle with very thick three-ply leashes, which gets overtaken by the weaving and is left two picks behind, after which the rod is withdrawn and the upstanding loops cut along the whole length,

---

[1] A like form of spool is found on the Sermata loom already mentioned.   Note, p. 68.

with a resultant pile. The rug on this loom was about three feet (or 1 m.) long, and several are made at intervals on one warp laying and beaming. When I pur-

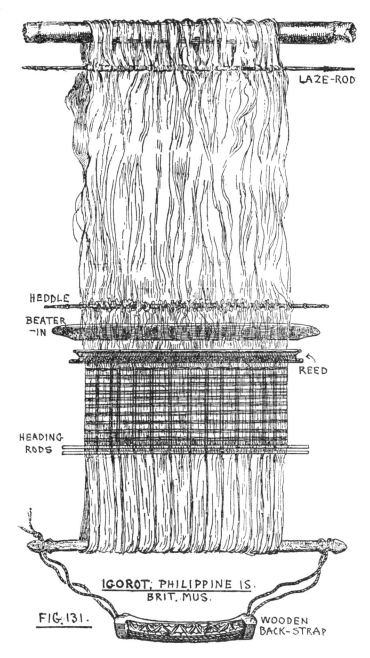

LAZE-ROD

HEDDLE

BEATER
—IN

REED

HEADING
RODS

IGOROT, PHILIPPINE IS.
BRIT. MUS.

FIG. 131.

WOODEN
BACK-STRAP

chased this specimen the heavy beater-in was not included in the sale, as I was told it was an heirloom without which the weaveress could not work, and a replica

74

was of no use to her as it did not and could not possess the qualities of the original. I had to content myself with the replica, and concluded it to be a case of weavers' ritual.

The Bhotiya loom is evidently the same as that described by Moorcroft and Trebeck as being in use among the Northern Ladakis.[1] The Igorot and Ilanun looms are a step in advance of the Iban and Dusun and Bhotiya looms in so far that they possess reeds.

An Igorot loom in the British Museum, obtained from Mount Isarog, Luzon, by Jagor (see Fig. 131) consists of a breast beam, two heading rods, one " single " heddle, a beater-in, two laze rods, a warp beam, four spools, and a wooden back strap or yoke. Length from beam to beam inclusive 42 inches (or 1·07 m.); width of web 15 inches (or 38 cm.). The warp, which is continuous, consists of a fine non-spun fibre (? *musa*), and so does the weft. There are 62 picks to the inch (or 24·4 to the cm.), and 28 warp (sisters) to the inch (or 11 to the cm.). In the web there is a wider space between every two warp threads than between every two picks, the picks being all equidistant. As in the Bhotiya and Ilanun looms the warp is wound round one of the laze rods (see Figs. 128 and 134). The pattern, an Oxford shirting design, is obtained by means of dark blue warp and weft at regular intervals. The spools are thin pieces of cane of varying lengths, viz., 38·5, 42·5, 44·5 and 52 cm. long respectively, that is to say they extend for the full width of the web and over ; three of them have form Ab1 and one approximating form Ab3. The heddle rod and laze rod ends are curiously pointed, like a round spear head. The heddle leashes are continuous, alternate, overlapping, and consist of strong doubled fibre.

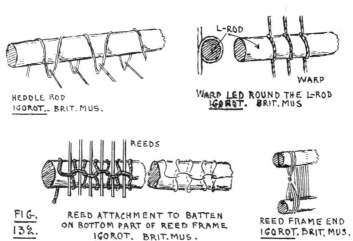

HEDDLE ROD
IGOROT.. BRIT. MUS.

L-ROD

WARP

WARP LED ROUND THE L-ROD
IGOROT. BRIT. MUS

REEDS

FIG.
132.

REED ATTACHMENT TO BATTEN
ON BOTTOM PART OF REED FRAME
IGOROT. BRIT. MUS.

REED FRAME END
IGOROT. BRIT. MUS.

The reed frame consists of two pieces of cane—a top piece and a bottom piece ; The teeth are of fine cane whose ends fit into a grove in the bottom piece, where

---

[1] *Travels in the Himalayan Provinces of Hindustan* . . . 1819-1825, London, 1841. ii, pp. 72-74.

they are fastened in position by means of some strongly twisted fibre which passes between every one of them, *i.e.*, through the dents, and round one or two slips of cane, placed on either side along the groove. The upper ends of the canes fit loosely *without* any tying up into the upper part of the frame, which has been split in two to receive them.

A loom from Sangir Island (Fig. 132A) between Celebes and Mindanao, obtained by the British Museum in 1872 (M. Steller), is similar to the Igorot loom. The

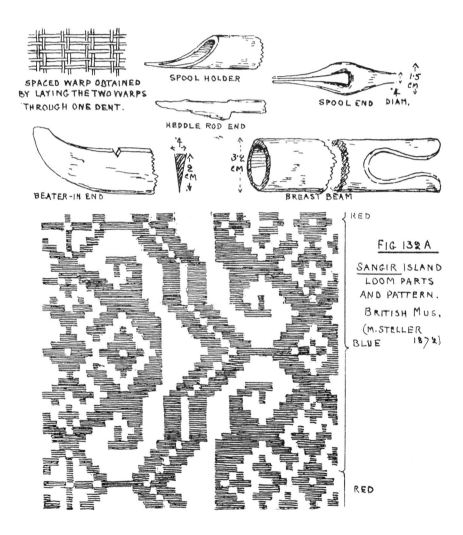

SPACED WARP OBTAINED
BY LAYING THE TWO WARPS
THROUGH ONE DENT.

SPOOL HOLDER

SPOOL END   DIAM.

1·5 cm

HEDDLE ROD END

BEATER-IN END

2 cm

3·2 cm

BREAST BEAM

RED

FIG 132 A

SANGIR ISLAND
LOOM PARTS
AND PATTERN.

BRITISH MUS.
(M. STELLER
1872)
BLUE

RED

particulars are as follows : length, beam to beam inclusive, 27 inches (or 68·5 cm.); width of web, 8¾ inches (or 22 cm.); 42 warp to the inch (or 16·5 to the cm.); 40 picks to the inch (or 15·7 to the cm.). The whole fabric is of non-spun fibre. The warp is made to keep in pairs by passing two of them through one dent. A piece of non-spun plaited fibre about 5 mm. broad appears to have been used as a back strap. There are two heddles with non-spun, continuous, alternate overlapping leashes. Non-spun leashes are rare. There is an elaborate brocaded pattern woven through the web in broad bands of blue and red alternately, the bands being of varying width of 3½ inches (or 8·9 cm.). Besides the fine small reed fixed top and bottom with fairly stiff canes and quite rigid, there is also a small light beater-in. There are three spools one each for the red, blue and buff weft, and as in the Ancient Peruvian loom (Fig. 40) and the Okale loom and the Borneo loom (Fig. 125), this loom is pro-vided with pattern laze rods.

The curious fact about these looms is that in addition to the reed they are furnished with a wooden sword beater-in as well.[1] Regarding this co-existence of reed and beater-in on one and the same loom, Meyer and Richter[2] say that " strictly speaking where the loom has been enriched by a reed the beater-in is superfluous, in the same way as our looms possess a reed, but no beater-in. The latter has been retained as a survival in order to give the reed efficient support (*festen Rueckhalt zu geben*) and to serve at the same time as a beater-in as before, which means that the beater-in was partly at least put to a new use." They say also that we must have more definite information as to the local use of the beater-in, on looms provided with a reed, in various parts of the Archipelago before we can adopt a definite con-clusion on the point. Failing the advent of such information I offer the following explanation :—The canes of the reed are not fastened to the upper bar of the reed frame (and the same absence of top fastening occurs in the Ilanun loom about to be described) and as a result when hard pressed these canes come away from the top bar, which necessitates the retention of the wooden beater-in with the object of its performing its work as before. But owing to the presence of the reed the beater-in cannot do the whole of the work it did before, and instead of assisting the reed and being thus put to a new use, the reed takes some of its work from the beater-in. When the canes of the reed are fixed top and bottom they have sufficient rigidity to beat-in, and seem then also made stouter and the wooden beater-in being no longer necessary gets gradually discarded. The co-existence of these two tools on one and the same loom therefore indicates a transition state, in which the primary use of the reed appears to be that of a warp spacer, before the discovery was made that it could be used as a beater-in as well. I do not think the absence of top fastenings on the reed is a sign of decadence, for the reason that the tool is not likely to have come into use full-blown, but by degrees and as a warp spacer form at first.

---

[1] C. M. Pleyte (*De Inlandsche Nijverheid in West Java*, Batavia, 1912, pl. viii) figures a loom from Zuid-Banten, with both reed and sword (beater-in). The details are not very clear, but if the reed is as flimsy as it looks the surviving presence of the more primitive beater-in is accounted for.

[2] *Webgeraet aus dem Ostindischen Archipele mit besonderer Ruecksicht auf Gorontales in Nord Celebes*, Ethnographische Miszellen. Dresden Museum, ii, No. 6, p. 47.

As a matter of fact, J. A. Loeber[1] says : " In Borneo there is to be found a very primitive form of this reed. It looks like a rake without the handle and betrays its purpose without any doubt." He tells us specimens from South and East Borneo can be seen in the Ethnographical Museum, Leiden, and one from Borneo in the Grassi Museum, Leipzig. He gives an illustration of such a warp spacer which is reproduced in Fig. 133 as nearly as possible, but his illustration is far too minute—

FIG. 133.

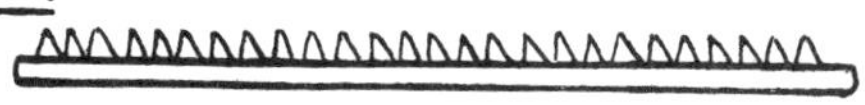

WARP SPACER ON A BORNEO LOOM, FROM J.A. LOEBER JR: WEVEN IN NEDERLANDSCH-INDIE. AMSTERDAM. 1903. P.30.

it is only 23 mm. long—for us to do more than to agree that it does represent a warp spacer, and to add that probably several warp threads pass through one dent, instead of every warp thread having a dent to itself. Whether the dents are produced by cutting notches in the stick or by the insertion of pegs is not clear from the illustration, but as Loeber says it looks like a rake, pegs must be inferred.

In the Cambridge Museum of Archæology and Ethnology there is an Ilanun loom from the Tampassuk district of British North Borneo, brought home in 1915 by Ivor H. N. Evans (Figs. 134, 135 and 136). It consists of breast beam, reed, two " single " heddles, two laze rods, warp beam, back strap, beater-in, and seven spools. Length, beam to beam inclusive, 25 inches (or 64 cm.) ; width of web, 33⅞ inches (or 86 cm.); 42 warp to the inch (or 16·5 to the cm.), all single ; 42 picks to the inch (or 16·5 to the cm.), all treble (" sisters "). The breast beam is 50 inches (or 1·27 m.) long, and more or less square, 4 by 3·8 cm.; the warp beam is 5 cm. square. The heddle rod and leashes are similar to those on the Igorot loom (Fig. 132). One laze rod is 3 cm. in diameter, the other is 1.4 cm. in diameter. The back strap (Fig. 134) consists of a piece of raw hide on the outside ; the inside or concave surface is covered with red cotton cloth, and this again is covered with a piece of green hide with a pattern cut out of it like fretwork.

There is one large transverse spool and six small spools (Fig. 136) for carrying the embroidery weft in mauve, orange, yellow, red, green and white. The warp laying repeats itself after every sixth thread. As the embroidery runs for every two and every four threads of warp (equals 6 threads) there is a correspondence between the warp laying and the brocading, from which one may conclude that the laying is intended as a guide to the brocading.

[1] *Het Weven in Nederlandsch-Indie Bull. Kolonial Mus, te Haarlem*, No. 29, December, 1903, p. 30.

The reed is similar in principle to the Igorot reed, that is, the canes are not fastened at the top and are very fine, and perhaps on account of their fineness, or to compensate to some extent for their want of top-fastening, or perhaps even as

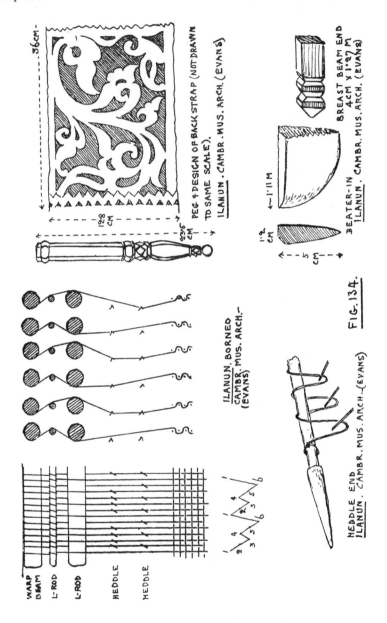

a step towards such fastening, the canes are loosely looped together for a distance of 3 cm. at one end and 7 cm. at the other end (Fig. 135). Altogether the reed frame is more elaborate in construction than the Igorot one, while the loose looping at the

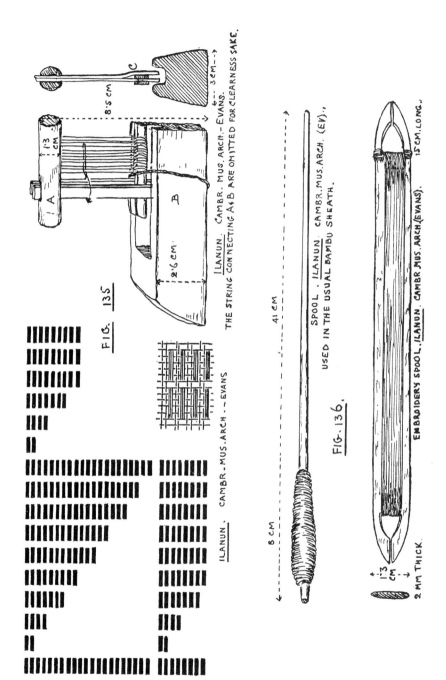

FIG. 135

ILANUN. CAMBR. MUS. ARCH. – EVANS.

THE STRING CONNECTING A & B ARE OMITTED FOR CLEARNESS SAKE.

ILANUN. CAMBR. MUS. ARCH. – EVANS

SPOOL. ILANUN. CAMBR. MUS. ARCH. (EV).

USED IN THE USUAL BAMBU SHEATH.

FIG. 136.

EMBROIDERY SPOOL. ILANUN. CAMBR. MUS. ARCH. (EVANS).

ends is a step in advance. With this we approach a completion of the chain of evidence of the evolution of the reed, for the next step is the making of a complete frame in which the canes are fastened top and bottom. To summarize it we have :—

(a) The Borneo warp spacer—a pegged rod allowing two or more warp threads to pass through every dent, with which the old sword beater-in is used quite independently.

(b) The pegged rod prevents entanglement, thereby assisting the progress of the work. This advantage is increased by having a dent for every warp, which in its turn necessitates finer pegs or canes, so that the increased number shall still fit into the same limited space. A finer yarn with more warps to the inch likewise necessitates finer pegs or canes.

(c) The finer canes are found to be too pliable, and require a top support, which is given by means of a groove in an upper bar.

(d) This step is followed by the perception that, in addition to its original function of a warp spacer, the now incipient reed frame could be made to act as a beater-in, with advantage to the evenness and closeness of the web. To do this the fine reeds must be fastened in the top bar as well as in the lower one, and as the frame becomes more rigid it adopts the secondary function, and the sword beater-in is ultimately discarded.

The *raddle*, or, as the Scotch call it, the *evener*, used as a warp spacer in laying out the warp, will probably have had a similar development to that of the reed.

The development thus described practically overcomes the difficulty referred to long ago by Tylor[1] of not being able to follow changes of one and the same people at different times, and satisfies the canon laid down by Karl Pearson[2]—that steps of sequence should be drawn from the usage of one tribe or group of tribes—for we see this evolution going on at this day among more or less allied peoples in one more or less restricted area.

The African herring-bone stick (Fig. 98) may have been evolved out of the old Roman spaced slot-rod found at Gurob, Egypt, which Flinders Petrie conjectures to have been a warp spacer. The distance in time from Egypt to Borneo is considerable, and if this Roman warp spacer has migrated eastwards it has had not only ample

---

[1] *Researches into the Early History of Mankind*, 2nd ed., London, 1878, p. 159 : " It happens unfortunately that but little evidence as to the early history of civilisation is to be got by direct observation—that is, by contrasting the condition of a low race at different times, so as to see whether its culture has altered in the meanwhile."

[2] *Grammar of Science*, 2nd ed., London, p. 359 : " To find sequences of fact—a growth of evolution expressible by a scientific law—we must follow the changes of one tribe or people at a time." His objection does not affect the question of the evolution of the shuttle either, for we find, as is the case with the reed, all its successive steps in a very circumscribed area.

time for its travels, but also ample time in which to alter or improve itself. That such alterations do take place we have plenty of evidence, and for our purposes we may cite as an example the case of the Santa Cruz loom. This, as I will show later on, in travelling from Indonesia to its present limit, has traversed almost as many miles as the Roman warp spacer must have traversed if it did go from Egypt to Indonesia. The Santa Cruz loom, in a probably much shorter space of time, has considerably altered, and, to some extent, improved itself, and we should expect some alteration in the Roman warp spacer. But the possible alteration from sawn slots to pegs is not much to find after a lapse of nearly two thousand years of travel, and as we see the evolution of the reed now going on in the East there is no need to search far afield for its origin. The fact that we have before us all the stages of this evolution in a restricted area makes it quite likely that the Borneo pegged stick warp spacer (Fig. 133) is indigenous to Indonesia, and this view is in accordance with the general evidence which tends to show that one birthplace of weaving was in this part of the world.[1]

An incomplete model of a Bugis loom in the Cambridge Musuem of Archæology and Ethnology, brought home by W. W. Skeat, belongs to the same form as that of the Igorot and Ilanun looms, being supplied with a " single " heddle and a reed. But the remarkable points are the method by which the warp is fixed on the breast beam and the curious grooving of the warp beam. The breast beam is made of two longitudinal blocks (Fig. 137), one being tongued and the other grooved longitudinally, which when fitted together hold the warp very securely. It cannot be revolved like a roller breast beam, because in section the two parts together are rectangular, that is to say, it has flat sides like a board, and can only be turned over

---

[1] Having used the words *sley* and *batten* (bottom of p. 57) without defining them, I have been asked by a vigilant student to explain their meaning. Mill managers, loom tuners and weavers, use the word sley and reed as synonymous terms for that collection of reed-canes (or reed wires in modern looms) which in their frame act as warp spacers and beaters-in, the workpeople generally using the word sley in preference to the word reed. They call the upper horizontal part of the reed frame the *hand-tree*, but Fox (*The Mechanism of Weaving*, London, 1911, p. 467) calls both the upper and lower part of this frame the *rib*, while others call these parts battens. Fox also gives the name sley to the shuttle box beam attached on power looms to the lower part of the frame (*op. cit.*, p. 326, Fig. 169). Definitions from practical men are not always alike. The beating-in of the weft " is performed by what is termed the lay, which carries the reed dividing the warp threads. The lay performs two distinct functions, the beating up of the weft and carrying the shuttle " (Thos. R. Ashenhurst, *A Practical Treatise on Weaving and Dyeing*, Huddersfield, 1893, 5th ed.). " The batten consists of two flat pieces of wood into which grooves are cut for the reed or sley, which is fixed in by iron or wooden pins, and is suspended from the capes of the loom." (Alf. Barlow, *The Principles of Weaving*, London, 1878, p. 62). Formally the whole reed frame, together with the two supporting side pieces and cross top piece, was known collectively as the batten; nowadays it is known as the *going part*. The word batten marked on the reed (Fig. 113) is used in the ordinary sense of a thin strip of wood, and in this instance to indicate that this portion of the frame is not the same as the heavy horizontal piece below it. The word side-battens mentioned on p. 58 is used also in the sense of a strip of wood.

from side to side so that the tension of the warp is not delicately adjusted, and must in most cases be too taut or too slack. This flat-sidedness constitutes a transition

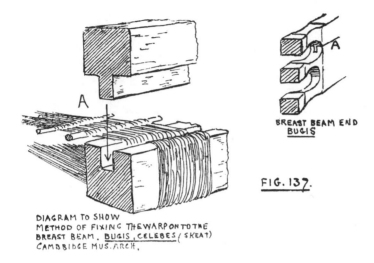

DIAGRAM TO SHOW
METHOD OF FIXING THE WARP ON TO THE
BREAST BEAM. BUGIS, CELEBES (SKEAT)
CAMBRIDGE MUS. ARCH.

BREAST BEAM END
BUGIS

FIG. 137.

towards the Malay and Cambodian loom, which we shall have to deal with directly. In principle it is somewhat similar to the very primitive African vertical mat loom beams (Fig. 50, etc.), which are likewise not revolvable, but have a groove cut length-

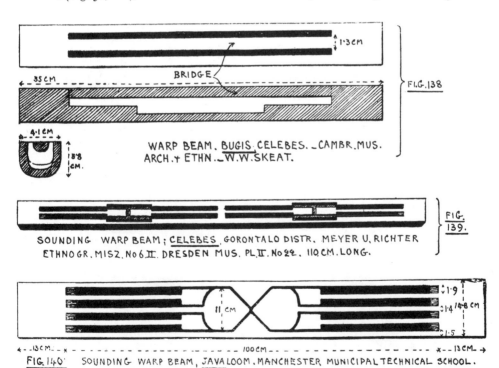

1·3 CM

35 CM   BRIDGE   FIG. 138

4·1 CM

3·8 CM

WARP BEAM. BUGIS CELEBES. CAMBR. MUS.
ARCH. + ETHN. W. W. SKEAT.

FIG. 139.

SOUNDING WARP BEAM; CELEBES, GORONTALO DISTR. MEYER U. RICHTER
ETHNOGR. MISZ. No6 II. DRESDEN MUS. PL. II. No 2½. 110 CM. LONG.

1·9
1·4 14·8 CM
11 CM
1·5

13 CM.   100 CM   13 CM.

FIG. 140 SOUNDING WARP BEAM, JAVA LOOM. MANCHESTER MUNICIPAL TECHNICAL SCHOOL.

wise, into which the warp is pressed and held down tightly by a rod pressed and tied down on top to prevent it from slipping. Meyer and Richter[1] illustrate a similar breast beam from Gorontales in Celebes.

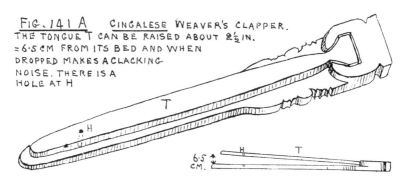

FIG. 141 A   CINGALESE WEAVER'S CLAPPER.
THE TONGUE T CAN BE RAISED ABOUT 2½ IN.
=6·5 CM FROM ITS BED AND WHEN
DROPPED MAKES A CLACKING
NOISE. THERE IS A
HOLE AT H

The warp beam of the Bugis model differs also in another respect from warp beams met with outside this region. It has two longitudinal grooves which join under the separating piece left, forming it into a longitudinal bridge (Fig. 138). A more advanced form of this grooving is illustrated by Meyer and Richter as belonging to the Celebes loom just referred to, in their Plate II, No. 22 (see Fig. 139). In the Gorontales illustration the resultant two bridges are cut up into four tongues enlarged at the loose end, and the authors speak of the whole as a *LaermVorrichtung*(signal-,rattle-,arrangement). A still more complicated form of this warp beam exists on a Javanese loom in the Manchester Municipal College of Technology (Fig. 140). It has six loose tongues, which vibrate with every movement on the loom, and strike against the back, making a rattling noise. I think the arrangement has something to do with the ritual of weaving, for Mr. W. Myers, M.Sc., lecturer in the textile testing department, informs me that when the loom was received the donor (whose

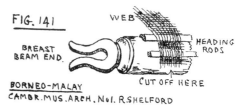

FIG. 141
WEB
BREAST
BEAM END.
BORNEO-MALAY
CAMBR. MUS. ARCH. No I. R.SHELFORD
HEADING
RODS
CUT OFF HERE

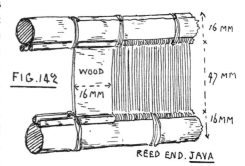

FIG. 142
WOOD
16 MM
16 MM
47 MM
16 MM
REED END. JAVA

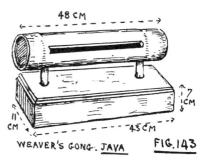

48 CM
WEAVER'S GONG. JAVA   FIG. 143

---

[1] *Op. cit.*, Pl. II, No. 1

name has been unfortunately forgotten) explained that every time a pick was made the Javanese weaver struck a bambu gong (Fig. 143), placed alongside, a sort of swishing blow with the sword beater-in. There are signs of wear on top of the gong, the striking of which can have nothing to do with the weaving, and which I would suggest is an act to propitiate or warn some spirit. In connection with this sounding warp beam there is an instrument bound up with some loom parts from Ceylon, in the British Museum, which may possibly likewise have something to do with weavers' magic (Fig. 141A). The tongue, which lies flat on its base, can be raised $2\frac{1}{2}$ inches (or 6 cm.) at the loose end, and makes a loud clacking noise when dropped. The hole H may have served to hold a knob.

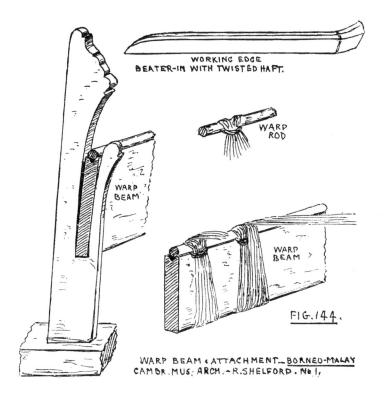

WORKING EDGE
BEATER-IN WITH TWISTED HAFT.

WARP ROD

WARP BEAM

WARP BEAM

FIG. 144.

WARP BEAM & ATTACHMENT.—BORNEO-MALAY
CAMBR. MUS. ARCH.—R. SHELFORD. No I.

We now come to the Malay, Javanese and Cambodian forms, a class of loom provided with a reed, and whose characteristic is the flat warp beam already referred to, combined with the rudiments of a loom frame.

The model of one of these is in the Cambridge Museum of Archæology and Ethnology. It was brought home by the late R. Shelford, who called it a "floor loom." It has the warp board fitted into a slot in the front edge of each of a couple of posts (Fig. 144). The warp board is provided with one "single" heddle and a shuttle of form Bb1. The canes in the reed frame are fastened both top and bottom, and not at bottom only. The sword beater-in has a bent haft somewhat like the handle of a kris[1]. The back strap is of wood. With the presence of the back strap and the single heddle the necessity for any loom frame has not yet become apparent, although the two warp posts form a beginning.

A similar form of loom is illustrated in Fig. 145, copied from that of a Bali weaver by Nieuwenhuis, in the periodical *Nederlandsch-Indie*. In this the warp board posts or supports are slotted from the top down the middle and not at the front edge. In another model (Fig. 146), also brought home by Shelford, the warp beam fits into a pair of posts swung from the top of the loom frame, which look as though they had originally been on the ground, as shown in Fig. 144. In a Pahang loom model given me by Leonard Wray, and at present in Bankfield Museum, these

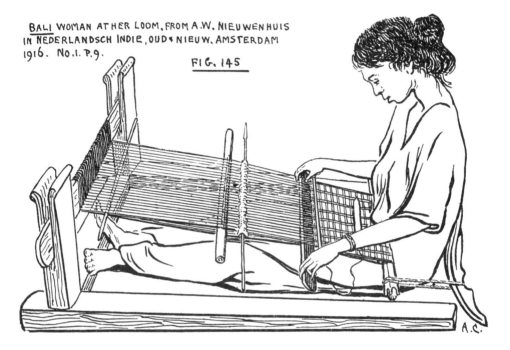

BALI WOMAN AT HER LOOM, FROM A.W. NIEUWENHUIS IN NEDERLANDSCH INDIE, OUD & NIEUW, AMSTERDAM 1916. NO.1. P.9.

FIG. 145

hanging supports have become elongated arms provided with oblong openings at their ends (Fig. 148) into which the warp board fits. This arrangement *looks* very like that of the " going part " of our hand looms.

A still further development is to be seen in a Kelantan loom mode brought home by W. W. Skeat, now in the Cambridge Museum of Archæology and Ethnology, in which the arms or side battens have disappeared altogether, leaving only the ends, furnished with the warp beam openings, which are held up by cord (Figs. 147 and 149). It will be noticed in one of the Malay and Bali arrangements (Figs. 146

[1] A model Malay loom in the British Museum from the Rani of Sarawak has a similar beater-in.

and 147), that the warp emerges, so to speak, from the lower edge of the warp board, while in another Malay loom (Fig. 144) the warp comes away from the upper edge.

In Raffles' *History of Java*, Plates, 1844, pl. IX, we are given an illustration of a loom, native name tenunan, from that island, which, while lacking somewhat in clear detail of the parts, gives us an intelligible idea of the whole (Fig. 150). On the other hand, Meyer and Richter give us fairly clear details of the parts of one loom, but curiously enough omit any illustration of the loom as a whole (even omitting details of the warp-board supports), so essential for arriving at a correct notion of what the loom is like. In this respect the full-sized loom from Java in the Manchester

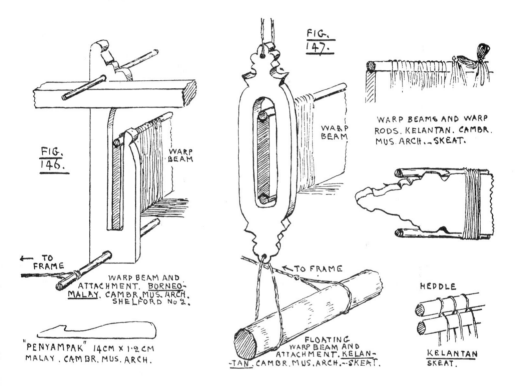

FIG. 147.

FIG. 146.

WARP BEAM

WARP BEAM

WARP BEAMS AND WARP RODS. KELANTAN. CAMBR. MUS. ARCH.—SKEAT.

← TO FRAME

WARP BEAM AND ATTACHMENT. BORNEO-MALAY. CAMBR. MUS. ARCH. SHELFORD No 2.

←TO FRAME

HEDDLE

"PENYAMPAK" 14CM X 1·2 CM MALAY. CAMBR. MUS. ARCH.

FLOATING WARP BEAM AND ATTACHMENT. KELAN--TAN. CAMBR. MUS. ARCH.—SKEAT.

KELANTAN SKEAT.

Municipal College of Technology, already referred to, may be examined with advantage. Set up it is very similar to the one depicted by Raffles. The approximate length of warp is about 100 feet (or 30 metres). It consists of the warp board, already described, supported on slotted posts, a breast beam, reed, " single " heddle laze-rods, shed sticks and wooden back strap,[1] temple wanting (Fig. 151). There are 82 warp (" single ") to the inch (or 32·3 to the cm.) and 49 picks (threefold) to the inch (or 19·3 to the cm.). Although a primitive loom, the work is equal in every way to the best that can be produced on any loom, the selvedge is excellent and the

[1] Of the same shape as that of the model Malay loom in the British Museum, referred to in note, p. 86.

88

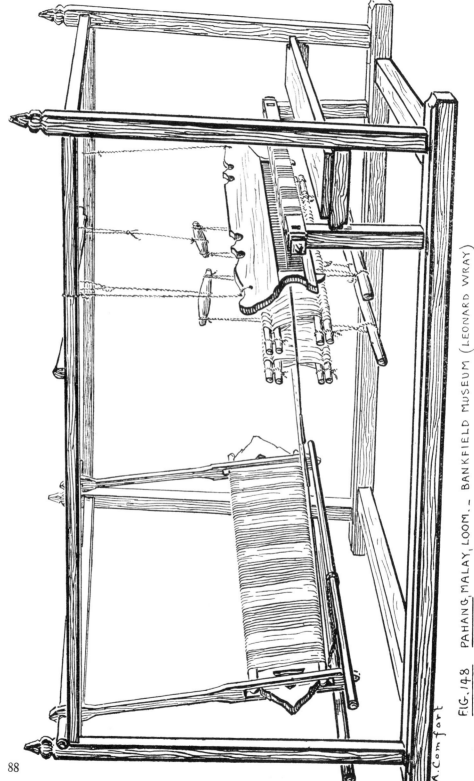

FIG. 148   PAHANG, MALAY, LOOM. – BANKFIELD MUSEUM (LEONARD WRAY)

A. Comfort

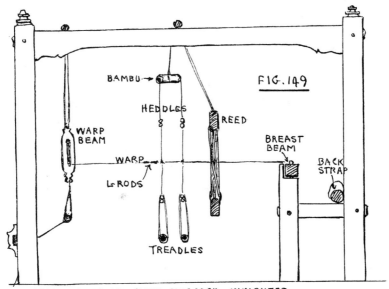

BAMBU →

HEDDLES

REED

WARP BEAM

BREAST BEAM

WARP

BACK STRAP

L-RODS

FIG. 149

TREADLES

KELANTAN, MALAY. — CAMBR. MUS. ARCH. — W.W. SKEAT.

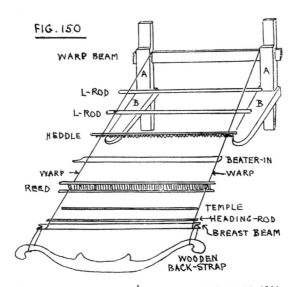

FIG. 150

WARP BEAM

A

A

L-ROD

B

B

L-ROD

HEDDLE

BEATER-IN

WARP

WARP

REED

TEMPLE

HEADING-ROD

BREAST BEAM

WOODEN BACK-STRAP

JAVA LOOM — RAFFLES' HIST. OF JAVA. PLATES. 1844
PL. IX. — FOR THE SAKE OF CLEARNESS ALL BUT THE
SELVEDGE WARPS HAVE BEEN OMITTED (THE NOMEN-
-CLATURE IS MY OWN).
AA WARP BEAM SUPPORTING POSTS SET IN THE GROUND
UP TO THE BASE OF BRACKETS BB.

web fine and even throughout. Apart from the figured pattern in the cloth itself, the brocade of gold thread—gold (?) tape wound round a two-fold yellow yarn—

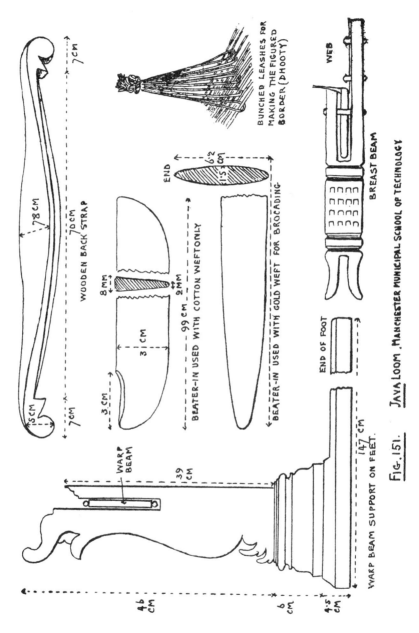

FIG. 151. JAVA LOOM, MANCHESTER MUNICIPAL SCHOOL OF TECHNOLOGY

necessitates a second set of heddles, and for making the dhooty (figured border) there is what is known as a dhooty bobby, a set of heddle leashes bunched, but without any rod, as in the African mat loom (Fig. 59) and the African cotton loom

(Fig. 87).[1] Separate hard wood polished sword beaters-in are used for the figured pattern and for the brocade pattern. The reed is well and neatly made, the canes are fastened top and bottom, and show considerable elasticity. The laze rod is of hard palm rind with the ends spear-shaped as in the Ilanun loom (Fig. 134). The shuttles belong to type Bb1, the gold-thread spool being topped with a carved knob.[2]

A flat warp board similar to those above described, but placed horizontally instead of vertically, is met with in Cambodia and is illustrated by J. S. Black[3] without any explanation in the text. I have reproduced it in Fig. 152 and the reproduction is, I hope, accurate in general, but owing to the smallness of the original the details cannot be correctly given. The points in this illustration which strike one immediately are the *downward* slope of the warp *away from* the weaver and the more or less flat position of the warp board, this board or beam being apparently not supported on posts, but fastened in position by means of cordage. It is probable that the flat outwardly sloping position of the warp board indicates the original position of this class of warp beams, such position being the least developed. The provision of double heddles and treadles renders some sort of framework necessary, in fact they are the cause of the existence of the framework, which appears to be made up of two distinct parts, viz. (1) the portion supporting the heddles and reed in which is placed the weaver's seat, and (2) the portion which supports the warp board or beam. Apparently these two portions are quite distinct, but have come together, forming perhaps the origin of the loom frame.

---

[1] Absence of rod is also mentioned by Harrison, 20, *op. cit.*, p. 48, with regard to the Lengua, South American, loom.

[2] Wm. Marsden mentions two forms of loom in Sumatra (*History of Sumatra*, London, 1783, p. 148), but neither of his descriptions is clear. He says : " Some of their work is very fine and the patterns prettily fancied. Their loom or apparatus for weaving (*tunnone*) is extremely defective, and renders their progress tedious. One end of the warp being made fast to a frame, the whole is kept tight, and the web stretched out by means of a species of yoke, which fastens behind the body, as the person weaving sits down. Every second of the longitudinal threads passes separately through a set of reeds like the teeth of a comb, and the alternate one through another set. These are forced home at each return of the shuttle, rendering the warp close and even. The alternate threads of the warp cross each other, up and down, to admit the shuttle, not from the extremities, as in our loom, nor effected by the feet, but by turning edgeways two flat sticks which pass through. The shuttle, *toorah*, is a hollow reed, about 16 inches long, generally ornamented on the outside and closed at one end, having in it a small bit of stick, on which is rolled the woof or shoot. The silk clouts have usually a gold head. They use sometimes another kind of loom, still more simple than this, being no more than a frame in which the warp is fixed, and the woof darned with a long small pointed shuttle. They make use of a machine for spinning the cotton very like ours. The women are expert at embroidery, the gold and silver thread for which is procured from China, as well as their needles. For common work their thread is the *poolay* before-mentioned, or filaments of the *pesang* (*musa*)."

[3] " A Journey round Siam," *Geographical Journal*, VIII, 1896, p. 435.

It seems that the formation of the complete loom frame out of two independent portions is due to the considerable development of the warp board supports, which

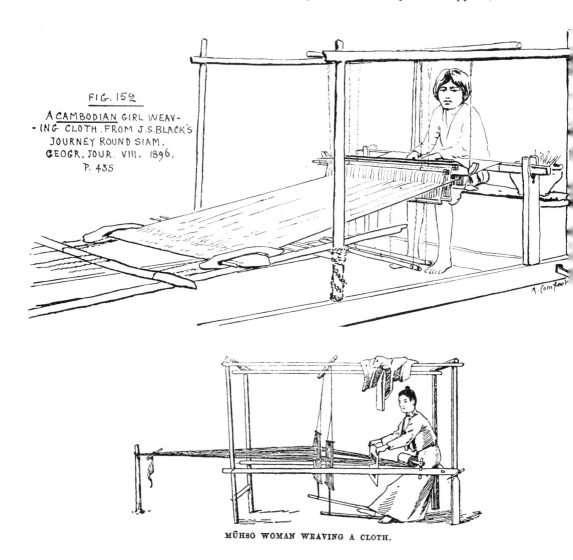

FIG. 152

A CAMBODIAN GIRL WEAV-
-ING CLOTH .FROM J.S.BLACK'S
JOURNEY ROUND SIAM.
GEOGR. JOUR. VIII. 1896.
P. 435

MŪHSO WOMAN WEAVING A CLOTH.

FIG.153. COLONEL R.G.WOODTHORPE : SOME ACCOUNT
OF THE SHANS. JOUR. ANTHR. INST 1896 XXVI. P.20.

we do not find outside Indonesia. The illustration (Fig. 153) of a Mühso loom as given by Colonel R. G. Woodthorpe[1] may at first sight appear to controvert the

[1] " Some Account of the Shans and Hill Tribes of the States on the Mekong," *Jour. Anthrop. Inst.*, 1896, p. 20.

fact that the frame is made up of two distinct portions, but the appearance of this Mühso loom indicates rather artificial than natural growth.   That is to say the growth has been due to exotic influences.   Thus the warp post points to a survival of the Pacific type of loom, and the free reed to a period previous to that of the frame which would be adopted with the double heddles and treadles when the latter were copied from the Chinese.   We have something similar in the Ashanti loom (Fig. 107), where a heavy stone, placed at some distance from the frame, serves as a warp beam ; but, as has been pointed out, this loom grew up under exotic (European) influence.

KOREAN WEAVER.   FROM A COLLECTION OF NAT-
-IVE DRAWINGS OBTAINED
BY H.M.BECHER ,ABOUT
1850.
BRITISH MUSEUM.

FIG.154.

WARP BEAM FROM
CAVENDISH'S KOREA
LONDON 1894. COL'D
PLATE P.P. 52-53.

FIG 154 A.

This flat warp beam is found also in Japan and Korea (Figs. 154, 154A), being, however, modified in both countries to the extent that the centre portion is cut away until the tool looks like a short square bladed canoe paddle.   The object served by the flatness of the beam is not very clear.   To a people devoid of mechanical genius, who are unable to make use of a round beam because they evidently could not invent a cross-head to prevent unwinding, a flat board would be sufficiently heavy to keep its position, and that position would be assured somewhat by a slope, as shown in the illustration of the Cambodian loom (Fig. 152).   On the other hand, with this sort of beam the tension is not so easily regulated.   On all looms provided with more or less squared, instead of round, beams the beams can only be turned the exact distance of the centre of one of the four sides to the centre of the next side, there being no intermediate shades of distance to get the exact amount of tautness required in the warp.   This want of means of adjustment may be the explanation of the survival of the back strap with complete loom frame as seen in the Kelantan loom (Fig. 149).

There are a few small points to call attention to in the loom frames.   The Pahang

93

Malay loom (Fig. 148) is supplied with the usual complete harness forwork ing the heddles (whipple trees and not pulleys being in use); the treadles consist of two pieces of bambu placed transversely to the warp, forming a convenient foothold. In the Cambodian loom (Fig. 152) the harness appears not to be so far advanced as

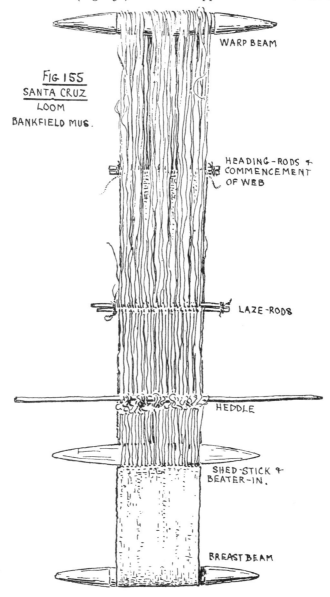

FIG 155
SANTA CRUZ
LOOM
BANKFIELD MUS.

WARP BEAM

HEADING-RODS &
COMMENCEMENT
OF WEB

LAZE-RODS

HEDDLE

SHED-STICK &
BEATER-IN.

BREAST BEAM

yet, and unfortunately is not as clearly shown in the original as could be wished, but the indication is to a loose cord passing from one heddle over cross pieces on top of the loom frame, and down to the other heddle. The treadles appear to be similar to those in Fig. 148, and in the Kelantan loom (Fig. 149) the upper cord joining the

94

heddles passes through the natural hollow of a piece of bambu supported on the top of the frame.

Belonging to the Indonesian series is the well-known Santa Cruz form of loom, chiefly remarkable for the long distance it has wandered away from what was presumably its original home. The loom is, as we shall see, not by any means limited to the island of Santa Cruz, but is found on the route to this group from the eastern outskirts of Indonesia.

The specimen in Bankfield Museum (Fig. 155), which is quite typical, consists of breast and warp beams, two heading rods, four laze-rods, one " single " heddle, sword beater-in and spool. The heading and laze-rods are narrow strips of cane ¼ inch (or ·6 cm.) across the flats ; the heddle rod is of wood 22¾ inches (or 63 cm.)

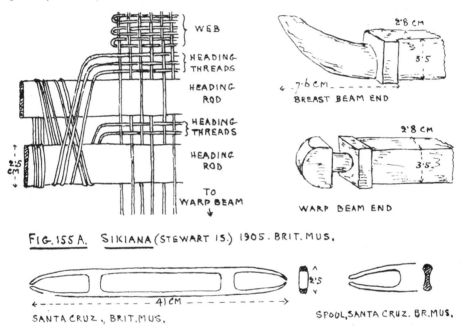

FIG. 155 A. SIKIANA (STEWART IS.) 1905. BRIT. MUS.

SANTA CRUZ., BRIT. MUS.

SPOOL, SANTA CRUZ. BR. MUS.

iong, the leashes being made of twisted fibre and are continuous, alternate overlapping. The beater-in is of hard wood fairly well polished, flat and oval and tapering from the middle to a fine edge all round, measuring ¼ inch (or ·6 cm.) in thickness, 1— inch (or 4·6 cm.) in breadth, and 15½ inches (or 38·4 cm.) in length. The warp consists of split non-spun filament, the end of one filament being knotted on to the end of the next, and so on, by means of which any length is obtainable. The weft is also of non-spun filament, said to be banana fibre. In the process of weaving both warp and weft—although in many parts the former is still splitting— are all single ; that is to say, we do not find two or more filaments acting as one warp as is the case with the Kwa Ibo and Ba-Pindi, African, looms. At the start the first pick is a piece of twisted fibre acting almost as a heading rod. In the use of non-spun filament, both as weft and warp, this loom shares the peculiarity with Ainu, Igorot and African looms, etc. The spool is form Ad1.

95

A somewhat modified form is that of the Sikiana loom in the British Museum, obtained in the year 1905. The fabric is of non-spun fibre, with a good selvedge and an Oxford shirting pattern obtained by means of white warp and red and blue weft, the bulk of the warp and weft being of the natural buff colour. The noticeable parts are the heading arrangement (Fig. 155A), which somewhat unusual form

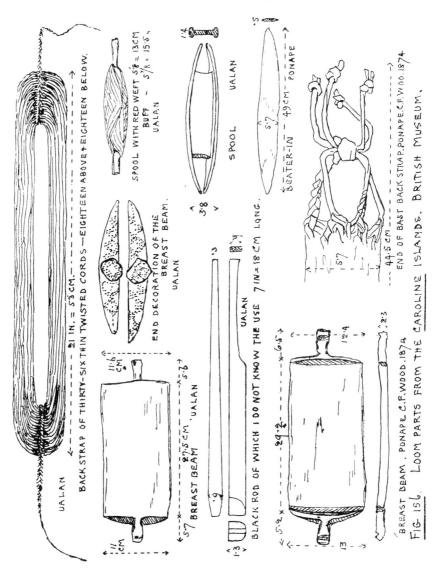

Fig. 156. Loom parts from the Caroline Islands. British Museum.

may possibly be due to difficulty in fixing up smooth non-spun filament and the perforated spool. A similar perforated spool accompanies a Santa Cruz loom obtained in 1891 from A. Lister Kaye. There are other specimens of Santa Cruz looms in the British Museum which do not call for special attention, except that on the last-mentioned loom there is a leash cord winder, form Ba, and that although the handle of the beater-in is broken, it may be like the handle in Figs. 165 or 167.

We now turn to the Caroline Islanders' loom. It is, unfortunately, the case that we are rarely favoured by travellers with any particulars of the preparation or laying of the warp which precedes the beaming or putting the warp on to the loom. In part this is due to the fact that in very primitive weaving the warp is laid on the beams of the loom as soon as there is a sufficiency of yarn. In the Caroline Islands, however, the method of laying the warp is so noticeable that the process has been recorded by several travellers. The first known description of the process is illustrated in a coloured plate in the atlas to Duperrey's *Voyage of the "Coquille,"* Paris, 1836, in which can be seen the warping bench and its peculiar grid. To explain the method of working it will be as well to examine first a piece of the fine matwork, made with non-spun fibre, which has been laid on this warping bench by means of the grid. The beautiful piece of work in question was obtained by C. F. Wood in Ualan and brought to the British Museum as far back as 1874.

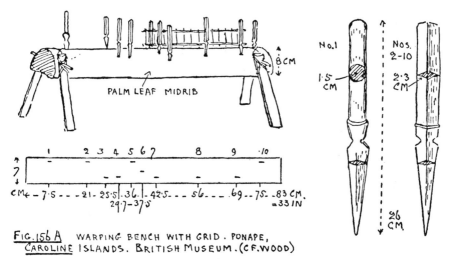

FIG. 156 A WARPING BENCH WITH GRID . PONAPE, CAROLINE ISLANDS. BRITISH MUSEUM.(C.F.WOOD)

The piece (Fig. 156B) may be divided up into the following sections (omitting the inch length of heading), viz.: I, $15\frac{1}{2}$ to 16 inches (or 41 cm.) long, with a red warp and alternate 2 picks of brown and 2 picks of buff throughout. II, 37 inches (or 94 cm.) long, in which the warp arrangement is 2 red, 2 buff, 2 red, 12 buff, repeated 18 times for the width of the web ; for a length of 33 inches (or 84 cm.) the picks are 4 buff and 2 brown repeat, and for a length of 4 inches the picks are all buff. III, 10 inches (or 25·5 cm.) long, the warp all buff with five patterns obtained by means of brown and red weft (Fig. 156C). The sections I, II and III are joined together by a simple knotting of the corresponding warp threads, and in doing this the worker has not been very careful or skilful enough to make the joint tally with a pick. That is to say, the joint passes askew along the picks over as much as $\frac{1}{2}$ inch (or 1·25 cm.) out of the parallel ; this may, perhaps, be due to the difficulty of making so many ties side by side, but the

97

difficulty can be overcome as shown at the end, for the joint there does run parallel with the pick. In another beautiful piece of similar work a belt formerly belonging to a Ponape chief, named Ometha, there are five warp-jointings, of which only one is perfect.

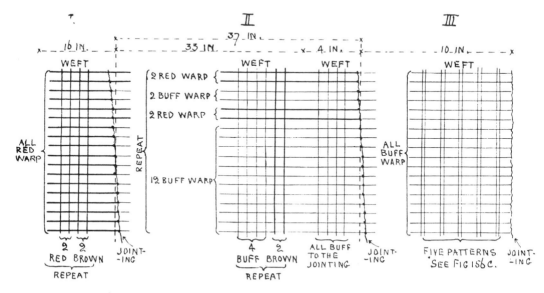

FIG. 156 B. DIAGRAM TO SHOW THE WARP SECTIONS ON THE PONAPE BELT, BRIT. MUS.

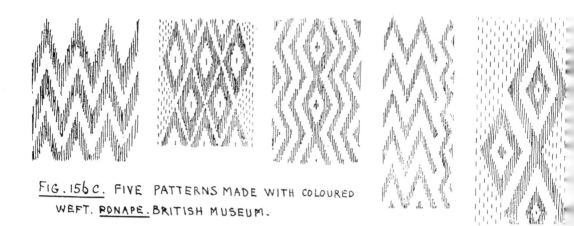

FIG. 156 C. FIVE PATTERNS MADE WITH COLOURED WEFT. PONAPE. BRITISH MUSEUM.

It is in order to regulate the length of the sections I, II and III, or as many more as may be necessary, to give the worker the measured length of warp required for the pattern essentially produced by the warp and not by the weft, that the grid comes into use, the pegs on the bench being for the usual warping purposes. The

Coquille grid has sixteen spaces between the bars of the grid from end to end, and so has the British Museum specimen. Finsch's illustration[1] shows ten spaces ; in the Hernsheim illustration[2] the particulars are too small to be enumerated. The Coquille bench has seven flat pegs and one round one. Finsch shows seven pegs, and the British Museum specimen has one round peg and nine flat (diamond section) pegs. The Hernsheim illustration shows six pegs.

My explanation does not tally with that of Finsch, who says : " Wie die Pfloecke die Laenge des ganzen Gewebes angeben, so das Heck die Laenge der gemusterten Endkante desselben, waehrend die Querstaebe des Heck wiederum die Laenge der einzelnen Querstreifen des Musters bestimmen (In the same way as the pegs indicate the length of the whole fabric, the grid gives the length of the patterns, while the bars of the grid settle the lengths of the individual patterns)." If the grid indicates the length of the patterned piece, *i.e.*, a section like I, II or II, a separate grid would be required for every section. It is the grid's bars, or two ends as the case may be, which regulate the warp lengths of the sections, which work is necessarily done on the warping bench. There is no necessity for, nor possibility of, regulating the length of the individual patterns like I 1, 2, 3, 4 and 5, because this is not done on the warping machine but on the loom by means of a greater or smaller number of picks.

The particulars of a loom parts from Ualan (Strong's Island) in the British Museum are as follows : beam to beam (warp beam missing), 37 inches (or 94 cm.); width of web, $6\frac{7}{8}$ inches (or 17·5 cm.). The breast beams, which have an elliptical section, are provided with lugs at both ends. The ends of a painted one are peculiarly decorated apparently by filling up small holes with lime arranged in triangular patterns. One back strap is made of eighteen strands of twisted fibre, 27 inches (or 68·5 cm.) long from head to head ; another back strap is $17\frac{1}{2}$ inches long by $2\frac{1}{4}$ inches wide (or 44·5 by 5·7 cm.), and is made of bast, with the ends plaited into the necessary loops for beam attachment. There are two small spools, form Ba, one filled with black and the other with red fibre, which, like the warp, is non-spun ; also a spool, type Ad1, body painted red and horns black. The Coquille illustration shows two spools of different shapes, one of which may possibly be similar to the New Britain modification (Fig. 161). It is accompanied by a short piece of hard wood, the functions of which are not clear to me and which may not belong to the loom. There is a much worn beater-in which is similar to the one in the Coquille illustration.

According to Kubary,[3] as regards the looms of the Ruk and Mortlock Islands

[1] *Ethnologische Erfahrungen*, Vienna, 1893, p. 220.

[2] *Sued-See Erinnerungen*, Berlin, 1883, p. 44.

[3] Schmeltz and Krause, *Die Ethnographisch-Anthropologische Abtheilung des Museum Godefroy in Hamburg*, Hamburg, 1881, p. 346.

the warp beam is fixed on to two upright posts. On St. Mathias' Island, north-west of New Hanover, according to R. Parkinson,[1] the weavers; who are women, in order to keep the warp taut, press their feet against the warp beam and their backs against the back strap, from which we may infer that on this island no up-right posts are used on to which to attach the warp beam. On p. 548, however, the same traveller illustrates a man seated at a loom, the warp beam of which is fastened to a standing tree trunk. Florence Coombe[2] gives an illustration (Fig. 157) of a man at Santa Cruz engaged at a loom, the warp beam of which is fastened to two upright tree trunks. One must conclude that, as among the Ainu, the warp was stretched by means of the feet in some localities and by means of fixed posts

FIG. 157

SANTA CRUZ ISLAND
WEAVER AT HIS LOOM.
FLORENCE COOMBE'S
ISLANDS OF ENCHANTMENT.
LONDON. 1911. FACING P. 183.

in others. But whether the warp beam is tied to uprights or is kept in position by the feet why in the specimens in the Cambridge Museum of Archæology and Ethnology, in the Norwich Castle Museum, in the Brighton Museum, in the Imperial Institute (which comes from Rotuma), and in all the specimens in the British Museum, should one cone-shaped end of the beam be more or less pointed and the other end have, as it were, its point bashed in or roughly flattened (Figs. 158, 166 and 168), as if the point had been hammered more or less flat? In the Bankfield Museum specimen the bashing is not so marked, and one beam is shorter than the other, their lengths being 16 and 17 inches (or 41 and 43 cm.) respectively. While

[1] *Dreissig Jahre in der Sued-See*, Stuttgart, 1907, p. 334.
[2] *Islands of Enchantment*, London, 1911, plate facing p. 183.

the ends are tapered more or less to a point as just mentioned, the bodies of the beams are not cylindrical, but of a slightly rounded square in section,[1] and this holds good of three of the other specimens referred to, the exception being the Rotumah warp beam, the section of which is rectangular. On the other hand, the New Britain specimen has both beams circular in section (Fig. 159), the breast beam having a neck to which to attach the back strap, the warp beam having a shoulder only. Schmeltz and Krause[2] speak of beams in the Lukunor group (Mortlock Island) as "about 97 cm. long, 16 cm. wide and 2 cm. thick" (38 inches by $6\frac{1}{3}$ by $\frac{3}{4}$), hence they must be rectangular in section, or board-shaped, and later on (p. 346) they quote Kubary, who says of the Nukuor (Carolines) warp beams : "The boards are made of hard wood, rectangular in shape, about 1 m. long, ·2 m. broad and ·03 m. thick (39 inches by $7\frac{3}{4}$ by $1\frac{1}{5}$)," the breast beam having lugs for the back strap, but the warp beam being without.

HAMMERED END OF BEAM, SANTA CRUZ

SECTION OF CENTRE OF BEAM SANTA CRUZ LOOM BANKFIELD MUS.

FIG. 158

Graebner[3] tells us of this Santa Cruz loom : " It is the old Indonesian loom as it exists in the western Carolines and has been only altered in somewhat essential points in the eastern part of the archipelago, in Kusaie and Ponape." The alterations he explains in a footnote as consisting in the board-like expansion of the beams and the introduction of the heddle frame. He continues : " All the same the original type on Truk-Mortlock (Central Carolines) has experienced one, even if trifling, change in the alteration of one of the laze-rods into a cylindrical stick round which the warp is wound."[4]

Graebner has, however, not realised that the flat beam is a characteristic of one of the Indonesian forms of loom, and that cylindrical laze-rods with the warp wound round once are found in the Bhotiya loom (Fig. 128) and Ilanun loom (Fig. 134), so that instead of these details showing modifications in some parts of the Carolines they are, in fact, survivals. Hence it would seem to be in the eastern portion of the Carolines, Ponape and Kusai, that the flat beam has held its own, and in the south

---

[1] In F. Graebner's *Voelkerkunde der Santa Cruz Inseln* (published in Foy's *Ethnologica*, i, Leipzig, 1909, p. 123, Fig. 53) the middle portions of both beams are seen to be more or less square in section.

[2] *Op. cit.*, p. 326.

[3] *Op. cit.*, p. 177.

[4] " Es ist der alte indonesische Webstuhl, wie er auch auf den West-Carolinen vorhanden und nur im oestlichen Teile des Archipels, auf Kuseie und wohl auch Ponape, in einigermassen wesentlichen Zuegen umgestaltet worden ist. Immerhin hat der urspruengliche Typus doch auch auf Truk-Mortlock eine, wenn auch geringe Umgestaltung erfahren, durch die Umwandlung des einen Faden-trenners in ein Rundholz, um das die Kettfaden herumgeschlungen werden." His footnote runs : "Durch brettfoermige Verbreitung der Spannhoelzer und durch Einfuehrung des Staebchenrostes."

central portion, Lukunor and Nukuor, the cylindrical laze-rod has held its own. It seems, further, that in leaving the Carolines the beams have lost their board-like character and have assumed the square shape preparatory to adopting the more practical cylindrical form.

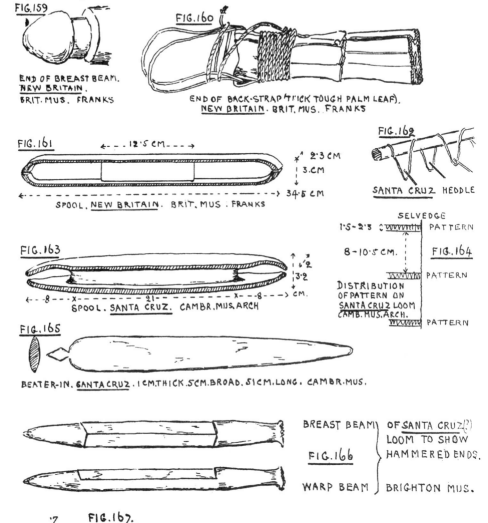

FIG.159

END OF BREAST BEAM.
NEW BRITAIN.
BRIT. MUS. FRANKS

FIG.160

END OF BACK-STRAP (THICK TOUGH PALM LEAF).
NEW BRITAIN. BRIT. MUS. FRANKS

FIG.161
←- - 12·5 CM - -→
x' 2·3 CM
3·CM
←- - - - - - - - - - - - - - - - - - - -→ 34·5 CM
SPOOL. NEW BRITAIN. BRIT. MUS. FRANKS

FIG.162
SANTA CRUZ HEDDLE

SELVEDGE
1·5-2·3 CM   PATTERN
8-10·5 CM.   FIG.164
                      PATTERN
DISTRIBUTION
OF PATTERN ON
SANTA CRUZ LOOM
CAMB. MUS. ARCH.   PATTERN

FIG.163
↑ 2
1·6·2
3·2
←- -8- - x - - - - - - 21- - - - - - - x - -8- - -→ CM.
SPOOL. SANTA CRUZ. CAMBR. MUS. ARCH.

FIG.165
BEATER-IN. SANTA CRUZ. 1 CM. THICK. 5 CM. BROAD. 51 CM. LONG. CAMBR. MUS.

BREAST BEAM │ OF SANTA CRUZ(?)
             │ LOOM TO SHOW
FIG.166   } HAMMERED ENDS.
             │
WARP BEAM   │ BRIGHTON MUS.

FIG.167.
·7
1·3
CM.
←- 82·CM.- - -→
SWORD BEATER-IN. NEW BRITAIN. BRIT. MUS. FRANKS.

In the following table I have grouped together for purposes of comparison particulars of dimensions and other details of the looms examined by me. As regards dimensions, warps and picks to the inch, etc., they are pretty much alike, except

102

the New Britain loom, which, already pointed out, has cylindrical beams and differs further in having the spool square nosed (Fig. 161) instead of pointed as in the other cases (Figs. 163, 169, 170). Schmeltz and Krause[1] describe in words some of the Santa Cruz spools in the Godefrey collections without giving any illustrations,

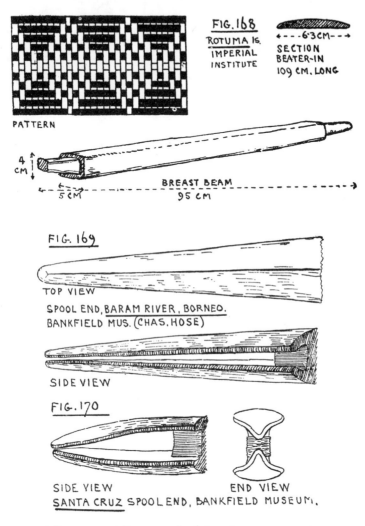

FIG. 168
ROTUMA Is.
IMPERIAL
INSTITUTE

SECTION
BEATER-IN
109 CM. LONG
← - - -6·3CM- - →

PATTERN

4 CM

5 CM

BREAST BEAM
95 CM

FIG. 169

TOP VIEW
SPOOL END, BARAM RIVER, BORNEO.
BANKFIELD MUS. (CHAS. HOSE)

SIDE VIEW

FIG. 170

SIDE VIEW   END VIEW
SANTA CRUZ SPOOL END, BANKFIELD MUSEUM.

which is not satisfactory, and I cannot find that they, or any other writers, refer to this form of spool end, although the Coquille illustration may possibly indicate a square-nosed spool end. The spool and the origin of the New Britain loom are out of the common and we need further information about it. It was purchased by the late Sir Augustus W. Franks for the British Museum, from a Norwegian captain.

[1] *Op. cit.*, pp. 326 and 345.

Origin of Loom	Museum where now placed	Length, Beam to Beam inclusive In.	Cm.	Width of Web per— In.	Cm.	No. of Warp to the In.	Cm.	No. of Picks to the In.	Cm.	Section of Warp Beam	Spool End	Warp and Weft	Heddle Leashes
Santa Cruz	Bankfield	43	109	6⅛	15·6	21	8·3	23	9	Square	Taper to a point.	All of non-spun filament.	Twisted filament, continuous, alternate, overlapping.
,,	Cambridge	27½	70	16½	42	34	13·4	16	6·3	,,	,, ,,		
,,	Norwich	38½	98	20⅜	52	35	13·8	21	8·3	,,	—		
,, (?)	Brighton	43⅓	110	35½	90	32¼	12·7	14½	5·7	Flat on 3 sides.	—		
Vera Cruz	British	19½	49·5	10¼	26	32	12·6	25	9·8	Square	Taper		
Sikiana	,,	49¾	126	21½	54·5	32	12·6	25	9·8	,,	—		
Santa Cruz	,,	30	76	7⅞	20	44	17·3	25	9·8	,,	Taper		
,,	,,	60	152·5	19¼	49	42	16·5	24	9·4	,,	—		
Rotuma	Imperial Institute.	52	132	—	—	25	9·8	19	7·5	Rectagular	—		
New Britain	British	28	71	25½	65	25	9·8	23	9	Circular	Flat-nosed		

The Rotuma specimen (Fig. 168), which is a typical Santa Cruz loom, except that the warp beam is rectangular, appears to indicate that the loom has travelled further west than has been suspected hitherto. Rotuma is an island about 480 miles almost due east of Tukopia, where Parkinson has reported[1] its existence, this island lying E.S.E. of Vanikoro, Santa Cruz group, and N.E. of Vanua Lava, Banks Island. That is to say the loom must have travelled twice the distance it did earlier in its migration from Pikiram (Greenwich) Island to Nuguria (Abgaris) Island. Nothing so very great. Unfortunately the authorities of the Imperial Institute can only say they received it labelled from Rotuma Island. One of the patterns (Fig. 168) is somewhat similar to those illustrated by Graebner, but nearly all his pattern illustrations are so small as to be almost useless, which is rather curious as he lays stress on the patterns as evidence of migration.

The patterns on these woven mats are arranged in more or less broad bands (Figs. 164 and 168) embroidered over the picks by means of banana fibre coloured black on one side only, that is they do not show on the wrong side. There is a loose specimen of this coloured fibre with the Norwich Castle Museum loom. Fringes also arranged in bands by insertion are common and on some there are loops arranged apparently for supporting the mats, which may have given origin to the fringe decoration. A specimen of the embroidery needle is shown in Fig. 171.

FIG. 171. EMBROIDERY NEEDLE, SANTA CRUZ, IN THE POSSESSION OF MR A. LISTER-KAYE, FROM EDGE-PARTINGTON'S ALBUM.

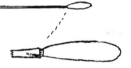

"LONG NEEDLE OF WOOD [38 IN = 96·5 CM] WITH LOOP OF COCO-NUT FIBRE PEGGED IN, USED FOR DRAWING THRO' THE BLACK THREAD WHICH FORMS THE PATTERN, THE FIBRE IS MADE FROM THE TRUNK OF THE BANANA SCRAPED DOWN AND BLEACHED."

The route by which this Indonesian loom reached so far east has been conjectured and studied by Codrington,[2] Parkinson, Graebner and others, from whose works I have prepared the accompanying map (Fig. 172). It would appear to have come via the Pelews and Carolines, and supposing it to have traversed the shortest route it would have come from Nukuor via Greenwich (Pikiram or Kapinga-marangi) Island and thence either to St. Mathias Island, and its neighbours Kerue and Squally Islands, or to Abgaris (Nuguria or Faed) Island, thence to Tauu (Mortlock) Island to Tasman Island (Nukumana atoll), to Ontong Java (Leventiua or Lord Howe's) Island, Sikiana (Stewart) Island to Santa Cruz group, thence to Banks Island, or

[1] *Op. cit.*, p. 343.
[2] *The Melanesians*, Oxford, 1891, pp. 20 and 316.

perhaps first to Tukopia, and thence to Banks Island. It will be seen that in the course of its migration it fringes the northern boundary of the Solomon Islands without establishing itself on them, a fact no doubt due to the ferocious nature of

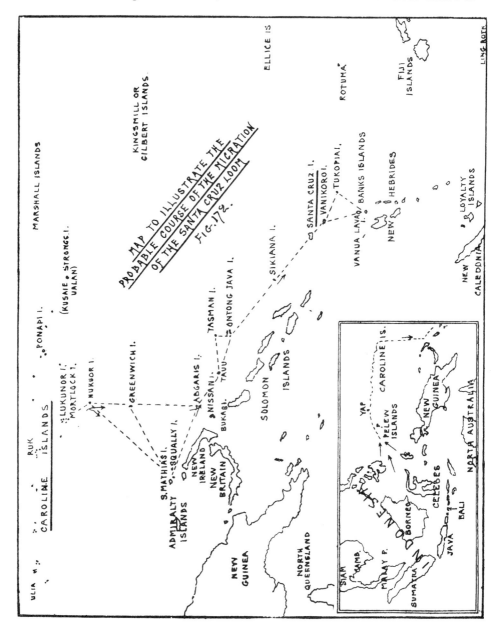

the natives there, who would be powerful and numerous enough to prevent the settlement on their shores of the higher civilized migrants who might have intoduced the loom. On the small outlying islands where the natives were fewer in number the immigrants would necessarily have more chance of settling and introducing their culture.

From all accounts in the Pelews, if it ever did exist there, and in most of the Carolines, as well as in the islands which formed the stepping-stones of its migration, the loom has now disappeared. Codrington[1] more than thirty years ago recognised that it had vanished from Banks Island, a disappearance which Rivers[2] ascribes to the loss of ritual essential to the working of the loom. Rivers' corollary that ritual was therefore an essential factor in primitive weaving may be supported, if support be necessary, by the conclusion one must draw from the Bhotiya webstress' refusal to work with any but a certain beater-in as mentioned on p. 74, and by the apparent ritual in use with the Java loom as referred to on p. 85.

[1] *Op. cit.*, p. 321.
[2] *History of Melanesian Society*, Cambridge, 1914, ii., p. 444.

7. THE SOLOMON ISLAND LOOM.

BEFORE discussing the Solomon Island loom, it may be as well to clear the atmosphere by calling attention to an article on an alleged South Sea loom, by A. Jannasch,[1] who gives an extraordinarily imaginative account of its development. Not understanding this, I wrote to F. von Luschan for enlightenment. He was kind enough to answer under date of 28th November, 1912, stating that, on writing to Jannasch, he only got an evasive reply, that Jannasch was probably mystified by some account of an Homeric or Greek loom, and that Jannasch is not to be taken seriously, and wound up by saying : " Anyhow, I think you need not trouble about his statement ; I am sure it is apocryphal, and I rather wonder that it has so long escaped the notice of ethnographers."

The Solomon Island loom was first described by Curt Danneil in a paper entitled " The Transition from Plaiting to Weaving."[2] He had found it on the island of Nissan (Sir Charles Hardy group). A similar loom, but from Buka island, exists in the Leiden Museum and there are four specimens, also from Buka, in the Dresden Museum. It is not clear whether the illustrations Danneil gives, reproduced in Figs. 173A and B, are those of the original article, but taking it for granted that he could not have produced such a delicate apparatus, we may accept the drawing as a representation of the native article. The loom is made up of a split piece of wood about 43 inches (or 110 cm.) long, the two halves tied together at the ends to prevent further splitting and kept asunder in the middle by two stays about $3\frac{1}{4}$ to 4 inches (or 8 to 10 cm.) long. A continuous yarn of bast is wound round that part of the frame which lies between the two stays, and this forms the warp; the pick is made in the usual way, apparently by means of the fingers and a needle. To raise the warp

[1] *Verh. Berl. Ges. f. Anthr.*, 1888, xx, pp. 90-91.

[2] " Der Uebergang vom Flechten zum Weben," *Archives Intern. d'Ethn.*, 1901, xiv, pp. 227-238.

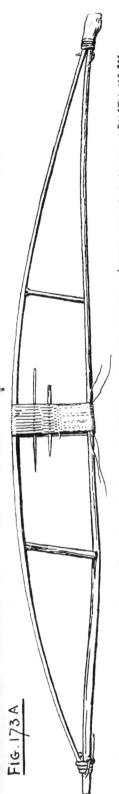

FIG. 173 A

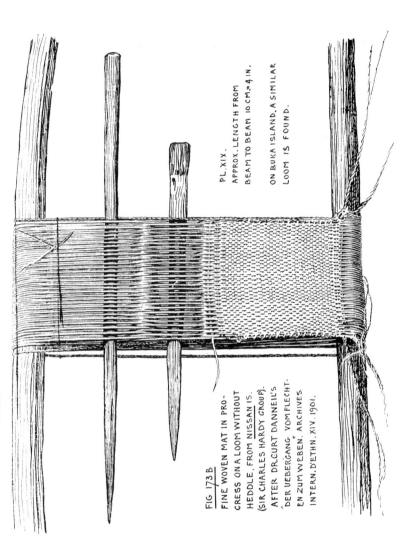

FINE MAT LOOM WITHOUT HEDDLE. NISSAN IS. FROM C. DANNEIL, ARCHIVES INTERN. D'ETHN. XIV. 1901. PL XIX., LENGTH 110 CM.

FIG 173 B
FINE WOVEN MAT IN PRO-
CRESS ON A LOOM WITHOUT
HEDDLE, FROM NISSAN IS.
(SIR CHARLES HARDY GROUP).
AFTER DR. CURT DANNEIL'S
"DER UEBERGANG VOM FLECHT-
EN ZUM WEBEN. ARCHIVES
INTERN. DETHN. XIV. 1901.

PL. XIX.
APPROX. LENGTH FROM
BEAM TO BEAM 10 CM.=4 IN.

ON BUKA ISLAND, A SIMILAR
LOOM IS FOUND.

and make the shed when necessary a pointed piece of wood is used. There is no mention nor indication of any heddle.

Danneil says of this loom[1]: " As it lies before us it represents for all time an original invention—an original transition from plaiting to weaving." He leads up to this claim by pointing out the difference between plaiting and weaving, saying that the first condition of weaving is the laying of the warp with the help of a warping frame (*Spannrahm*) and continues : " It was without doubt the nature of the material which put primitive man on to the idea to ' lay ' it and to construct a frame with that end in view, for fineness and want of stiffness made any material useless for free hand plaiting. It being necessary that one portion of the filaments should be ' laid ' once over *it resulted of itself* that another form of intercrossing of the filaments took place. Man already knew the material, either he had used it in making thread or had adopted it in a stiffer form (that is, not split up into such fine slips) for free-hand plaiting. With frame and warp primitive man had discovered the art of weaving, etc., etc." In all this there is no trace of any attempt to show how weaving arose out of plaiting or that it did so.[2] The connection is a close one, but as I have endeavoured to show later on, plaiting is not in the direct line of the evolution of weaving.

On the other hand, Meyer and Richter[3] aver that " this apparatus is no loom at all, as Danneil thinks, but a plaiting frame (*Der Apparat ist kein Webegestell wie er meint, sondern ein Geflechtrahmen*),[4] which opinion is apparently founded on the fact

---

[1] *Op. cit.*, p. 230.

[2] This want of demonstration on the part of Danneil in presenting his notion as to how weaving developed out of plaiting is on a par with Julius Lippert's presentation of his notion of the development of the shuttle. " The tedious passing of the weft with the fingers corresponded with the oldest art of sewing. The oldest stone needle was only an awl, with which a hole was made in order to put the thread through with the fingers. The more modern needle is, however, an awl which not only makes the hole but carries through the attached thread. Now in carrying this progress forward from plaiting (*Bandflechten*) to weaving *the shuttle developed itself*, the shuttle being nothing else than a needle fully specialised for this object. The completion consisting in the fact that it carried with it a lengthened thread wound round lengthwise in the same way as in our modern netting needles." (*Die Kulturgeschichte in einzelnen Hauptstucken.* Part I. Wohnung u. Kleidung, 8vo., Leipzig, 1885, p. 170.)

[3] *Op. cit.*, p. 61.

[4] Writers do not always discriminate in the use of the words weaving and plaiting—*Weberei* and *Flechten*. Buschan, in describing some plain weaving, gives an illustration of a loom on which such weaving is done and calls it a *Flecht-rahmen, i.e.*, a plaiting frame. (*Die Anfaenge u. Entwickelung der Weberei der Vorzeit. Verh. d. Berl. Ges. Anthr.* 1889.) W. H. Holmes, in referring to a pair of sandals which as he says " shows the method of *plaiting* practised by the ancient inhabitants of Kentucky," goes on to tell us that these sandals are " beautifully *woven*." Then he illustrates a " similar method of plaiting practised by the Lake Dwellers of Switzerland," and in the legend to the illustration calls it " braiding " (*Rep. Bureau of Ethnology*, 1882, p. 418). The italics are mine. The difference between weaving and plaiting has been explained in *Ancient Egyptian and Greek Looms*, p. 36. Similarly many writers speak of a spindle when a spool or bobbin is meant : the explanation may be that the article was once a spindle, but if its use is turned into that of a weft carrier it is no longer a spindle,

that it is not supplied with a heddle. But the correctness of the drawing being accepted, we have here a frame on which a web can be constructed by means of interlacing of one set of filaments at right angles to another set of filaments with the possibility of the attachment of a heddle, and it is this possibility which helps to confirm the fact that the apparatus, however primitive, is a weaver's loom.

Fine matwork made of delicate coloured strips of bast is one of the characteristic arts of the Solomon Islanders, and we find almost throughout their islands that it is largely in use as decoration for weapons such as spears, clubs, arrows and also for combs. The work is extemely beautiful and I very much doubt whether it has been surpassed anywhere, and this is especially the case with the ornamental head combs. Some years ago, in describing a few of the native weapons from these islands, I had occasion to remark : " It is curious to note that this matwork apparently all runs parallel with the outlines of the article ornamented, while in most cases in Borneo and wholly so far as I am aware in British Guiana,[1] the pattern is made to run diagonally across the article."[2] In other words, in the Solomon Islands we have to do with matwork, the basis of weaving, while elsewhere we have to deal with plaitwork. In so far as I can ascertain no one has yet described the method of

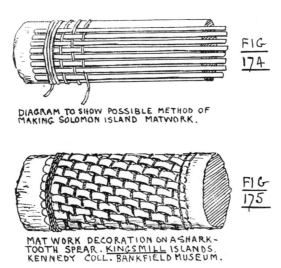

FIG 174

DIAGRAM TO SHOW POSSIBLE METHOD OF MAKING SOLOMON ISLAND MATWORK.

FIG 175

MATWORK DECORATION ON A SHARK-TOOTH SPEAR. KINGSMILL ISLANDS. KENNEDY COLL. BANKFIELD MUSEUM.

working, nor the seat of the manufacture, which still remains unknown,[3] at least, in so far as the beautiful coloured matwork combs are concerned.

An examination of the finished matwork on a flat club from Guadalcanar in

[1] Not wholly so in Brit. Guiana.
[2] " Spears and Other Articles from the Solomon Islands," *Archives Intern. d'Ethn.*, vi, 898, pp. 154-61.
[3] *Op. cit.*, p. 8.

the Kennedy Collection in Bankfield Museum tends to show that in its manufacture two methods are possible, but by both methods we get cylindrical or tubular or seamless garment weaving. One method is to wind a continuous filament spirally round the club, thus making it into a warp and then passing the other set of filaments at right angles through the warp as in making a pick. The other method is to tie the one end of a series of filaments side by side parallel with the outline lengthwise of the club and then wind a continuous filament spirally round the club in and out of the set of filaments as one would make a pick as indicated in the illustration, Fig. 174, a method which is accomplished in a cruder way in the Kingsmill Islands, as shown in Fig. 175.[1]

The same method—tubular weaving or seamless garment weaving[2]—appears to be followed in making the fine matwork covering of a Uganda child's cylindrical girdle, in making the coarse outer sheath of a British Guiana quiver, in making a small Andaman basket, and so on.[3] By this method a club can be covered with matwork from end to end as in this case, Fig. 174, for a length of 28½ inches (or 72·4 cm.). This is where, I think, the Nissan and Buka loom comes in. It comes in as an apparatus for weaving the matwork and has developed as a side issue to the Solomon Island tubular matwork ornamentation, the loom giving us as a product an enlargement of the club matwork with this difference, that what was originally the spiral continuous *weft* has become the spiral continuous *warp*. The loom described by Danneil and illustrated in Figs. 172A and B, is thus of local origin and has arrived at that stage where a heddle could be applied, but its development is now for ever arrested by the intrusion of the white man. Although it was present in close proximity to the Santa Cruz loom it evidently had nothing whatever to do with that exotic article.

8. A Lapp Woman's Belt Loom.

This little loom (Figs. 176, 177, 178) comes from the River Tana, Finmark, Norway, and is now in the Victoria and Albert Museum. The interesting point about the loom is the secondary heddle arrangement for weaving the floating pattern by means of the warp, a method rare in primitive looms, but of course common enough

---

[1] In Bankfield Museum there is a gourd stopper similarly decorated, obtained in New Guinea about 1886 by my brother, Dr. F. Norman Roth, which Dr. Haddon assigns to the Massim District.

[2] M. D. C. Crawford (*Peruvian Textiles, Anthrop. Papers Amer. Mus. Nat. Hist.*, Vol. xii, Part iii, New York, 1915, p. 95) says of tubular weaving that it " seems the most unlikely for the primitive craftsman to stumble upon," but here we have it in almost its very first stages among a very savage but artistic people.

[3] The examples quoted and others can be seen in Bankfield Museum.

in our manufactures. The length of warp as illustrated is 57 inches (or 145 cm.), and the width of the web is 1⅝ inches (or 4 cm.). The number of warp is 55 to the inch (or 22 to the cm.), and the number of picks to the inch is 22.5 (or 9 to the cm.).

FIG. 17b.

LAPP BRAID LOOM, FINMARK.

VIC. + ALB. MUS.

A

B KNOTTED END OF RED
WOOL PATTERN WARP

C CORD HEDDLE

B' RED WOOL
PATTERN WARP

BONE HEDDLE

It consists of eighty-eight warp threads, which are laid through a free rigid bone heddle, forty-four passing through the slots in the heddle and forty-four passing through small holes in the bars of the same, the shed being made by alternately

113

raising and lowering the·bone heddle. When the heddle is lowered or raised the warp threads passing through the holes are those which get lowered or raised, at the same time the threads passing through the slots practically remaining stationary. The movement is quite simple.

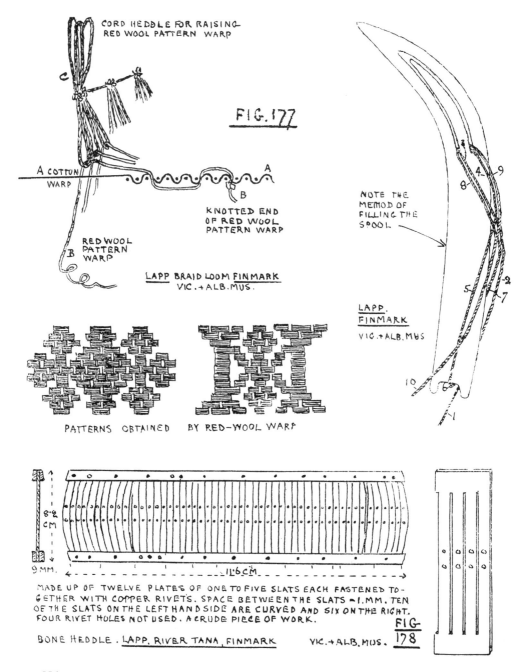

CORD HEDDLE FOR RAISING
RED WOOL PATTERN WARP

FIG. 177

A COTTON
WARP

KNOTTED END
OF RED WOOL
PATTERN WARP

RED WOOL
PATTERN
WARP

LAPP BRAID LOOM FINMARK
VIC.+ALB.MUS.

NOTE THE
METHOD OF
FILLING THE
SPOOL

LAPP.
FINMARK
VIC.+ALB.MUS

PATTERNS OBTAINED BY RED-WOOL WARP

8·2 CM

9 MM.

11.6 CM

MADE UP OF TWELVE PLATES OF ONE TO FIVE SLATS EACH FASTENED TO-GETHER WITH COPPER RIVETS. SPACE BETWEEN THE SLATS ~1.MM. TEN OF THE SLATS ON THE LEFT HAND SIDE ARE CURVED AND SIX ON THE RIGHT. FOUR RIVET HOLES NOT USED. A CRUDE PIECE OF WORK.

BONE HEDDLE. LAPP, RIVER TANA, FINMARK   VIC.+ALB.MUS.

FIG. 178

To obtain the pattern in this case, got by means of red spun wool, a series of eighty-seven red threads of equal length to the cotton warp are knotted under the warp at the breast beam end (Fig. 176B), and as soon as the three or four picks have been made, every red thread is drawn upwards separately between the warp threads until the knot stops its progress. Then, say as a start, a couple of picks are made and the loose ends of the red thread brought down between and below the warp. To facilitate this process a very primitive cord heddle is brought into use (Figs. 176C, 177C), every leash of which holds one red thread just as with ordinary primitive heddles. The leashes are bunched in threes and tied together at requisite intervals with special loops at each end of the row, apparently intended to be used for raising the red threads altogether. In working, every leash will be raised separately, or in threes, and, when the pick has been made, the red threads are pulled down underneath separately by the fingers. Naturally as the work progresses the cord heddle must be pushed further and further away. The pattern is worked on the wrong side, i.e., the pattern appears on the under surface while the work is in progress.

Otis T. Mason[1] has given us descriptions of the free rigid heddles in use among the Pueblo and other Indians and the white population of the United States, Germany, Finland, etc. In the Pueblo heddle the cross bars are tied on to the rectangular frame, but among Europeans, and also in Indonesia, the frame is carved out of one piece of wood. In the Lapp specimen (Fig. 178) the cross bars are cut out of twelve pieces of bone which are riveted on to a top and bottom bar.

There is no frame to the Lapp loom, otherwise apart from the warp pattern it belongs to the Norwegian type of belt loom in Bankfield Museum (Fig. 179), with a similar free rigid heddle. In the English eighteenth to nineteenth century ladies' ribbon loom (Fig. 180), instead of the rigid heddle being free it is fixed at the end of its frame box, and hence, as it cannot be raised or lowered to make the shed, the warp has to be raised and lowered instead ; but in this case the warp passing through the holes will remain stationary while the warp passing through the slots gets moved up and down. The same procedure is followed in the use of the rigid heddle when it stands by itself, as it still apparently exists in some parts of Germany and Indonesia.[2]

Specialisation in primitive looms, as in the above, is not uncommon, as we shall see in the next chapter.

---

[1] " A Primitive Frame for Weaving Narrow Fabrics," *Rep. U.S. Nat. Museum*, 1899, pp. 487-510.

[2] R. Stettiner : *Das Webebild in der Manesse Handschrift*, Berlin, 1911, p. 7; Meyer and Richter, *op. cit.*, Pl. II.

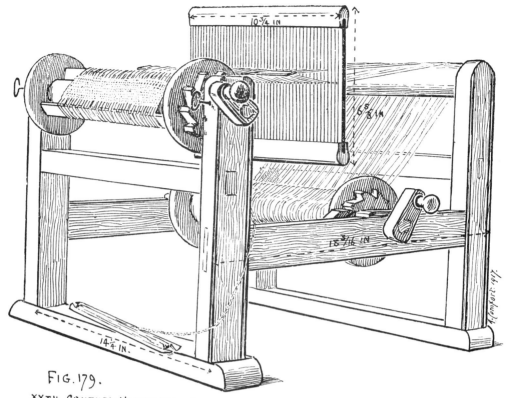

FIG. 179.

XX.TH. CENTURY <u>NORWEGIAN</u> LOOM WITH <u>FREE RIGID HEDDLE</u>. BANKFIELD MUS.

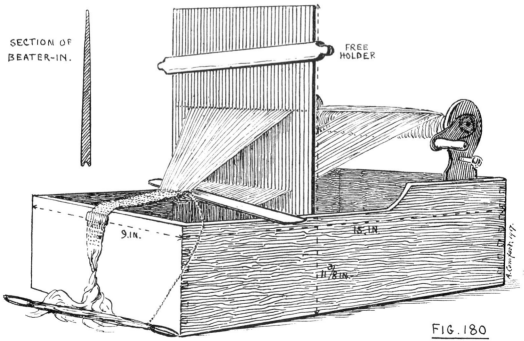

SECTION OF
BEATER-IN.

FREE
HOLDER

FIG. 180

LATE XVIII.TH. CENTURY ENGLISH LOOM WITH <u>FIXED RIGID HEDDLE</u>. BANKFIELD MUS.

116

9. ORIENTAL VERTICAL MAT LOOMS.

The Upright Oriental mat looms on which large and thick floor mats are made are of special interest because of the peculiar development of the beater-in, which consists of a heavy bar of wood with transverse slots for the warp threads to pass through. This development appears to be due to the springy nature of the material,

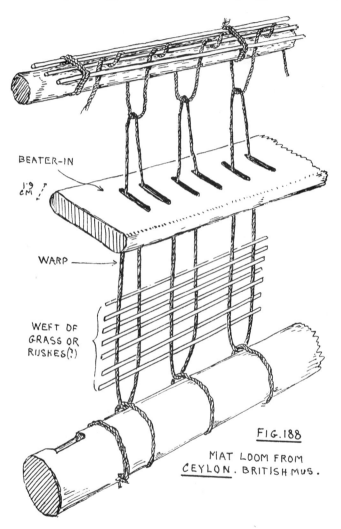

BEATER-IN

1·9 cm

WARP

WEFT OF GRASS OR RUSHES(?)

FIG. 188

MAT LOOM FROM CEYLON. BRITISH MUS.

straw, rushes, thick grass, etc., of which the weft is composed and which requires something heavy to hold it in position as the work proceeds.

In the specimen of the Ceylon mat loom (Fig. 188) in the British Museum, the beater-in is made out of one solid piece of wood of the following dimensions, 37 x 3 x ¾ inches (or 94 x 7·6 x 1·9 cm.). There are 112 slots for the warp to pass through ; the slots begin at a distance of about 2⅜ inches (or 6 cm.) from each end and are approximately $\frac{9}{32}$ inch (or 71 mm.) apart. In the specimen from Hong-kong (**Fig.** 189), at the Imperial Institute, the beater-in is more massive, to correspond with the heavy elaborate frame and thicker weft used, and is provided with special **handles** $5\frac{1}{2}$ inches (or 14 cm.) long ; the slots alternate in two lengths, the object of the longer ones is to allow more play and so obtain alternate long and short weft surface, as shown in Fig. 190. The action is clear enough and I am unable to follow Otis Mason when he says " the Chinese have a large block of wood with saw cuts inclined so as to throw the warp up and down in weaving the Canton matting,"[1] but there is no throwing the warp up and down, for it consists of rigid, strong yarn, as in ordinary looms. In this Hong-kong mat loom there are eighty-four warp threads in a mat-width of three feet.

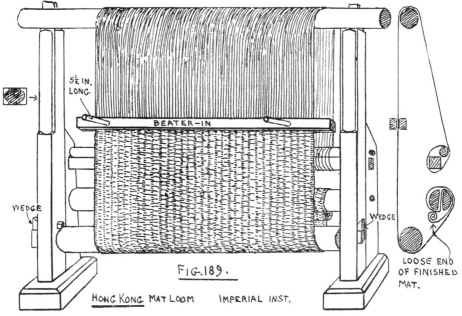

FIG. 189.

HONG KONG MAT LOOM   IMPERIAL INST.

WIDTH BETWEEN POSTS 8 FT. WIDTH OF MAT 3 FT.   84 WARP.

The mats obtained on both of these looms are true weaves and differ, therefore, from those made on the vertical mat-making frame of the Ainu. This consists of a ground beam and an upper beam supported by two uprights, the whole having the appearance of a rectangular frame, stood upright, resting on the ground-beam side. Two threads are fixed at intervals on the ground-beam opposite each other ; these threads are somewhat longer than the intended length of the mat and have each a stone fastened at the loose end. The work begins by placing rushes along the ground-beam between the opposing threads, raising these threads over the rushes, twisting them half round each other and then throwing them over the upper beam so that

---

[1] *Origins of Inventions*, London, 1895, p. 247.

one thread end with its stone hangs over one side and one thread end with its stone over the other. Then a second row of rushes is laid on top of the first row between the opposing threads and as before the threads are twisted over them and thrown over the upper bar, and so on—the twist always being made in the same direction. As the work proceeds and the mat is completed as far as the upper beam, it is rolled round the ground-beam, leaving a similar clear space as there was at first between the last or top row of rushes and upper beam to allow the work to be continued. By lengthening the threads the mat can be made of almost any length. Hitchcock's verbal description[1] not appearing, to me at least, as sufficiently explanatory, I had a frame made at Bankfield Museum according to the illustration he supplies us with, and have taken the above description from the actual working on this model.

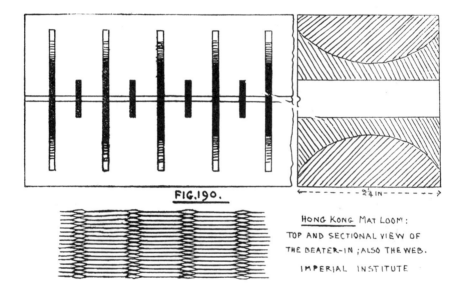

FIG.190.

HONG KONG MAT LOOM:
TOP AND SECTIONAL VIEW OF
THE BEATER-IN; ALSO THE WEB.
IMPERIAL INSTITUTE

In the Ainu mat frame the laying of the warp, if one can so call it, as the work proceeds is again probably due to the springy nature of the weft, which seems to require something more than mere interlacing to be kept in position. This something more is attained by twisting the warp threads after every piece of rush weft has been placed in position. With this method no beater-in is required. The loom and frame give a somewhat striking example of achieving the same result by different means. The Ainu frame appears to be the more primitive of the two and has differentiated at an earlier stage, but the mat-loom has probably an origin closely allied to that of the upright looms met with elsewhere. To get at the bottom of this we must hark back a bit.

---

[1] *Op. cit.*, p. 463.

In the Vatican library there exists an illustrated MS. book of Virgil's *Æneid*, of which photographic reproductions[1] have been made and distributed to various libraries in different parts of the world. The original is generally considered to be a production of about the fourth century A.D. On Fol. 38, Picture 39 gives a representation of the magic doings of Kirke and on the upper right-hand corner there is depicted a wooden structure (Fig. 191), which may be likened to a vertical loom. It consists of two uprights on feet connected by three equidistant horizontal bars with an irregular clear patch just above the lowest bar. The middle bar probably

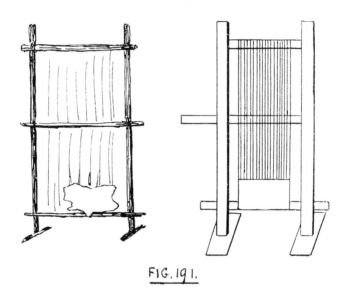

FIG. 191.

THE VIRGIL LOOM ACCORDING TO :-

FRAGMENTA ET PICTURAE CODICIS VATICANI. 3225. .... ROMAE. 1899. PICT. 39. FOL. 58

ANTIQUISSIMI VIRGILIANI CODICIS FRAGMENTA ET PICT- -URAE EX BIBLIOTHECA VATICA- -NA. ROMAE CIɔIɔCCXLI. P 129.

represents the heddle rod. The drawing of the structure is only about 24 mm. high, and this minuteness, together with the wear and tear of ages and the final photographic reprint, make it by no means a clear representation. A female figure (Kirke) standing to the left of the loom is depicted with her right hand on the junction of the heddle rod and upright post ; her left hand is probably also on the post lower down but not clearly shown. In neither hand does she appear to hold anything. Johannes Braunius[2] gave a very much larger illustration of this loom with the female on the right hand holding a wand in her right hand and showing a large

---

[1] *Fragmenta et Picturae Vergiliana Codicis Vaticani* 3225. Romae, 1899.

[2] *Vestibus Sacerdotum Hebraeorum*, 1680.

rectangular piece of cloth at the bottom of the loom. For a representation of this piece of cloth there is little warranty—for it is difficult at the present day to be certain what the white blotch was intended to represent in the original *pictura*. However, the Vatican published in 1741 an edition of the above-named copy of Virgil,[1] and in this Kirke and her loom are illustrated fairly distinctly, though on the same minute scale as the original (*see* Fig. 191). She is depicted not quite as in the original with her right hand on the heddle rod extension, while the left hand is not shown at all. In the meanwhile B. de Montfaucon[2] published a reverse of the illustration of the loom as it appears in Braunius, showing Kirke on the left again. Johannis Ciampini[3] follows Montfaucon almost to a line. Since then the illustration has been fancifully and thoughtlessly copied times out of number. But we have to come back to the point that this illustration probably represents a fourth century A.D. upright loom, in which the warp weights have already been replaced by a breast or cloth beam and the weaving begins from the bottom and not from the top. It is, in fact, an earlier form of the upright loom as we meet with it in the East, between Asia Minor and India, and also in Africa at the present day. Yates and Marindin[4] consider the making of the web to begin at the bottom as an anachronism, that is if we consider the period of Æneid's travels, but it really represents the artist's limited local knowledge of a loom in his days.

The loom referred to by Yates and Marindin is the well-known warp-weighted loom, a highly specialised form of which was depicted by Johannes Braunius, above referred to, over two hundred years ago (*see* Fig. 192). Both Bluemner and Marquardt condemn this as a piece of fiction, but give no reason for doing so. I have submitted the illustration to several practical weavers, and their opinion is that the working is quite feasible and to anyone who takes the trouble to examine the details of the illustration the feasability quickly becomes manifest. Montfaucon, in copying Braunius, gives an incorrect version of it and Johannis Ciampini has again apparently used Montfaucon's plate, reproducing the same mistakes both in essentials and in details. It has been re-illustrated many times until it has reached its final stage of degradation in an extraordinary work by Perry Walton.[5]

---

[1] *Antiquissimi Virgilium Codicis Fragmenta et Picturae ex Bibliotheca Vaticani*, p. 129.

[2] *L'Antiquite Expliquee.* Paris, 1719, Part iii. Pl. 195.

[3] *Romani Vetera Monimenti* Romæ, 1747, Pl. 35.

[4] Smith's *Dictionary of Greek and Roman Antiquities*, 3rd Ed., 1890.

[5] *The Story of the Textiles*, Boston, Mass., 1912. The adjective " extraordinary " has not been used inadvisedly. What is one to think of such statements as the following : " Fabrics dating back to a period thousands of years ago have been unearthed in England (p. 14)." " On the walls of Nineveh, Babylon, Thebes, and the ancient cities of Peru and Mexico, throughout most of the ruins of Assyria, Persia, Egypt, and among similar ruins of both North and South America, is depicted the whole process of the textile industry, from the raising of the sheep or growing of the flax to the spinning of the yarn and weaving of the fabrics " (p. 16).

FIG. 192

FROM JOHANNES BRAUNIUS
VESTIBUS SACERDOTUM
HEBRAEORUM.
1680.

P. Scoopendaal Sculptr

The following is the description of this loom as given by Braunius. It is worth reproducing, quite apart from the rarity of the book and its inaccessibility to the general public and even to students.

AAAA.—Loom, or ancient weaver's beam. An *upright* loom (Artemidorus, Bk. iii, Chap. 36). Perhaps called " jugum " by Ovid on account of its shape, which is not unlike a yoke. In what manner a yoke was constructed, and what was meant by " sent under the yoke," may be clearly seen from Cicero, *De Officiis*, Bk. iii, and Livius, Bk. iii, etc.

B.—Shirt, rounded and closed without seam ; " seamless " (ἄρραφος) as was the shirt of Christ (John, chap. xix). Otherwise " tunica recta." (Isidorus, Orig. Bk. xix, chap. xxii). This shirt is woven in an upward direction ; for the weaving begins from the topmost thread CC and gradually works down to D. (Herod., Bk. ii, Theophylactus " In Johannem," Festus Chrysostomos " In Johannem Homil." lxxxv. Isidorus Pelusiota, Epist. lxxiv, Bk. i), The shirt is rounded and closed from B to I; then, however, it is divided to D and E, as men's undergarments usually are to-day.

CC.—Threads, which are part of the weft (trama), but so prolonged beyond the body of the shirt that at last they can be made the warp (stamen) of the shirt-sleeves. When the finished shirt is taken off the loom, the threads CC are cut at the ends ; they are afterwards turned in, and finished of in the same way as BD.

DE.—Two warp-threads, of which D is the anterior, and E the posterior ; they are joined by one and the same thread to the weft, and plaited together ; " δυο ῥάκη συμβάλλειν," " duos pannos committere." (Chrysostomos " In Johannem Homil." lxxxv ; Theophanes Cerameus, " Homil. in Passion Domin." xxvii. Josephus, Bk. iii, chap. 8).

FF.—Weights with which the warp threads in this manner of weaving were weighted (Seneca, Epist. xc ; Pollux, Bk. vii, x).

G.—Spatha σπάθη, an instrument used for keeping the threads of the weft together (Seneca, Epist. xc ; Pollux, Bk. vii, chap. x).

H.—The woman-weaver, holding the spatha in her right hand for the purpose of bringing the weft together, by pushing the threads upward ; in the left hand she holds the weaver's shuttle. Moreover, she weaves standing, not sitting (Isidorus, " Orig." Bk. xix, chap. xxii. Servius Aen., Bk. vii ; Eustathius, " Ad Homer Odyss.," Bk. v ; Hesiod, " Ergon "; Artemidorus, Bk. iii, xxxvi). As she weaves she walks round in a circle ; for when she has passed the shuttle or weft through the web or threads D, she has to go round the whole loom, so

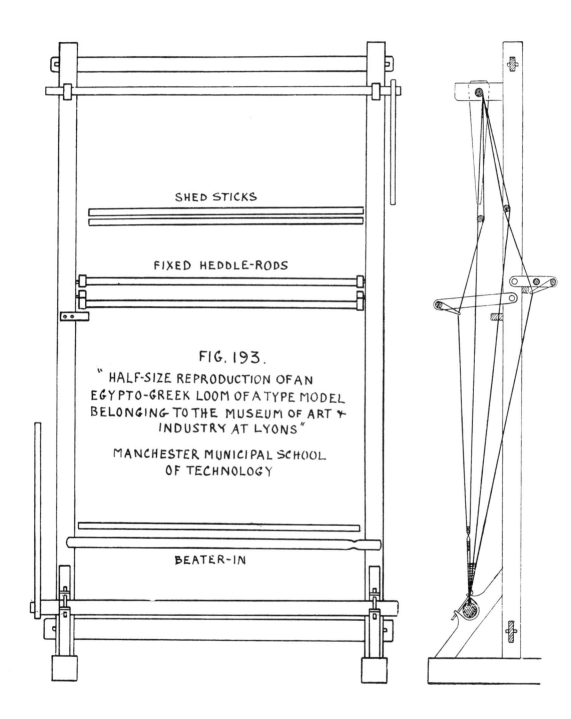

SHED STICKS

FIXED HEDDLE-RODS

FIG. 193.

" HALF-SIZE REPRODUCTION OF AN
EGYPTO-GREEK LOOM OF A TYPE MODEL
BELONGING TO THE MUSEUM OF ART &
INDUSTRY AT LYONS "

MANCHESTER MUNICIPAL SCHOOL
OF TECHNOLOGY

BEATER-IN

that she may pass the same shuttle and weft through the threads E, in order that the webs D and E may be woven together (Theophylactus, " In Johannem "; Virgil, " Aen.," Bk. vii ; Isidorus, " Orig." Bk. xix, chap. xxiv ; Artemidorus, Bk. iii, chap. xxxvi).[1]

The loom is one designed for making a seamless garment, and in fact produces what is called tubular weaving. That it has not survived is no doubt due to its complicated nature, coupled with the warp weight system. It remains, however, of considerable interest, inasmuch as the method of warp weighting depicted may perhaps indicate a transition from the use of simple warp weights to the adoption of a warp beam. Before proceeding further it may be as well to call attention to another form of tubular weaving as illustrated by a model in the Manchester Municipal College of Technology, of which the label reads " Half Size Reproduction of an Egypto-Greek Loom of a type model belonging to the Museum of Art and Industry of Lyons." The Textile Department cannot tell me anything as to its history, and I am unable to obtain particulars. The accompanying illustration (Fig. 193A) will explain its details and at the same time indicate that it partakes of the nature of a fixed heddle loom (although the heddles are not completely fixed) somewhat like the Aures loom (Fig. 91B), which may, to a limited extent, explain the name Egypto-Greek.

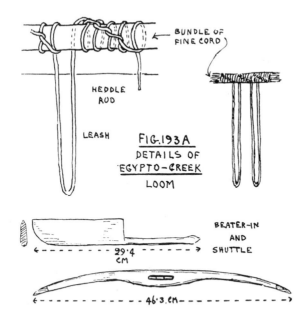

BUNDLE OF FINE CORD

HEDDLE ROD

LEASH

FIG. 193A
DETAILS OF
EGYPTO-GREEK
LOOM

29·4 CM

BEATER-IN AND SHUTTLE

46·3 CM

---

[1] For assistance in the translation of this description I am much indebted to my friend, Lieut. Arthur Redford, late Bradford scholar, Manchester University.

Reference has been made above to the upright looms found in Asia Minor, etc., which, like the upright looms in North Africa, are in all probability the immediate successors of the ancient warp-weighted loom. A few remarks on two of such looms may not be out of place.

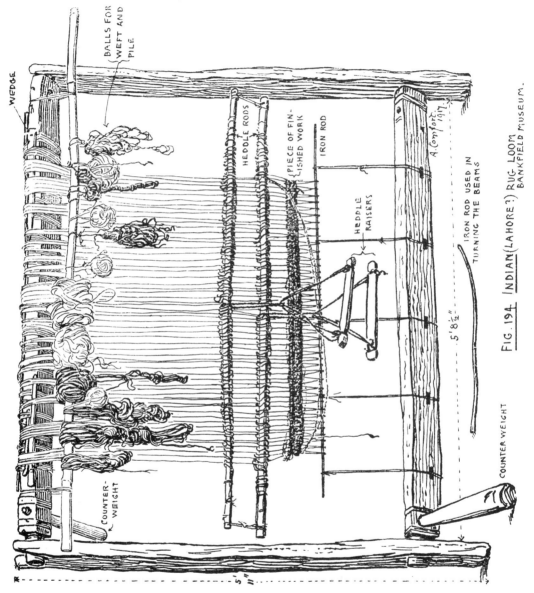

FIG. 194. INDIAN (LAHORE?) RUG LOOM. BANKFIELD MUSEUM.

The Bankfield specimen, said to come from Lahore, is depicted in Fig. 194. It is a rug loom, 71 inches (or 1·8 m.) high by 67 inches (or 1·7 m.) between the uprights. To a certain extent the warp is kept taut by means of heavy timber levers or counterweights as shown, the lower one of which, when in use, was apparently tied down to the ground.   To increase the tautness, but only in a very inefficient way, wedges are driven into the coils of warp on the upper beam.   On a bambu rod placed across the loom are hung variously coloured balls, with which to make the pile and weft, the threads being pulled out as required by the worker.   At the lower end the warp is attached to an iron rod, which in turn is attached to the lower beam by means of cords let into small rectangular holes cut into one edge of the beam.   The heddles are provided with raisers.   For every one row of pile there are three of weft.   The pile ends are cut level by means of a pair of shears which are provided with special lugs to keep them level when the loose ends of the pile are being trimmed.   The picks are driven home by means of a bent iron beater-in.   It is altogether a very crude loom.

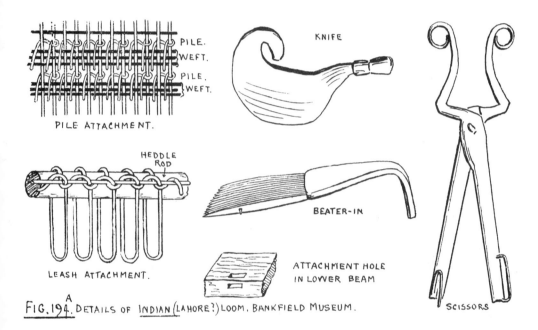

FIG. 194. DETAILS OF INDIAN (LAHORE?) LOOM, BANKFIELD MUSEUM.

But quite as crude is the rug loom illustrated by O. Benndorf[1], reproduced in Fig. 195.   Here the lower beam is fastened down by a cross bar passed through a hole at the end of the beam.   The beater-in is very crude, and is similar to one in the Victoria and Albert Museum (Fig. 195A) said to be Persian.

The frame of the two looms just described consists of two upright posts and two cross pieces which join the uprights at top and bottom respectively.   The frame of the Oriental mat loom with its specially developed beater-in belongs to this form. In the warp-weighted loom there is only one cross piece which joins the uprights at the top.   As incidentally mentioned when discussing Braunius' loom, there is an indication of a transition between these two looms, which consists in bunching the

[1] *Reisen in Lykien u. Karien,* 1884, p. 18.

lower warp ends to a loose rod, on to which one weight only is attached, which keeps all the threads taut. But there must have been an earlier or simpler frame than that of the warp-weighted loom. An example of this is the Kwakiutl loom, figured by

Fig. 12   Türkin am Webstuhl

FIG. 195. FROM O. BENNDORF'S REISEN IN LYKIEN U. KARIEN. 1884.

Mary L. Kissel,[1] or the Ojibway loom figured by M. D. C. Crawford.[2]   It consists of two uprights stuck into the ground about 2 feet apart and joined at the top by a piece of yarn, or perhaps originally sinew.   The weaving naturally proceeds

[1] *Aboriginal American Weaving*, Nat. Assoc. Cotton Manufacturers, Boston, Mass., 1910, p. 4, Fig. 1.

[2] *Amer. Museum Journ.*, Oct., 1916, p. 382.

downwards. On the Ojibway loom the cloth is apparently made in one piece. On the Kwakiutl loom the weaving is done at twice, that is to say, the cloth is woven for the full length of one half of the warp, and then the weaving continues or rather recommences on the top of the second half, and the two finished pieces are laced together at the adjoining edges. On the well-known Chilkat loom[1] the cloth is woven in several strips, instead of two only, and then joined up.

Besides the Kwakiutl loom, Miss Kissel illustrates[2] a similar frame to the above, but with a wooden cross-piece at the top, instead of a piece of string, on which mats are *plaited*. In Bankfield Museum there is a piece of plaited work of bison hair yarn given me several years ago by Miss M. A. Owen,[3] which has apparently been made on such a frame in narrow strips which have been laced together, and I have had a facsimile piece of plaitwork made on such a frame. In the Pitt-Rivers Collection, Oxford, there is a larger piece of the bison hair plaitwork which, until one examined the selvedge, has the appearance of diagonal weaving ! Advocates of the theory that weaving was evolved from plaiting would no doubt consider that these examples of primitive frames, so identical in construction on which both plaiting and weaving can be done, supports their theory. Both plaiting and weaving require some sort of simple framework support, so there is nothing in the coincidence. The presence of two sets of elements in weaving does not necessarily mean an advance over the one set of elements in plaiting. The initial step in plaiting, the selvedge, which is a *sine qua non* of plaiting, is a secondary matter in primitive weaving and has, as it were, to be undone or dropped or ignored if we are going to weave ; this would be a retrogressive step and places plaiting in the position of a side product rather than in the direct line of the evolution of weaving.

II. The Alleged " Weavers' Comb."[4]

In Figs. 181 and 182, outline illustrations are given of two of these tools now in the British Museum ; the larger one was found at Mortlake on Thames and the

---

[1] Emmons, *op. cit.*, p. 343.

[2] *Op. cit.*, p. 6.

[3] Author of " The Folklore of the Musquakie Indians," *Folklore Society's Journal*, 1904.

[4] Even if the Glastonbury and other similar tools were intended for beating-in the weft, and this is what is claimed as their function, it is a misnomer to call them " weavers' *combs*." The name comb implies an instrument for straightening or separating out any more or less tangled fibres by drawing it through the entanglement. In driving home the weft the action is not that of combing, but of a decided tapping or pressing down—there is no separating or straightening out of fibres, for this is not wanted, or if it were wanted it would be exceptional. When A. Barlow (*History of Weaving*, Lond. 1878, p. 58) wrote : " It is far from being uncommon for weavers at the present day to use a comb, especially when they have a sticky warp to weave, or a warp, that, owing to the felting property of the material, requires to be separated frequently," he was dealing with exceptional circumstances. A more appropriate designation of the tool would be a toothed beater-in.

smaller one in Kent's Cavern. Both are of bone and both are concavo-convex in cross section and in both the dents (spaces between the teeth) are of varying depth. The Mortlake specimen has fairly regular teeth of equal length ; in the Kent's Cavern

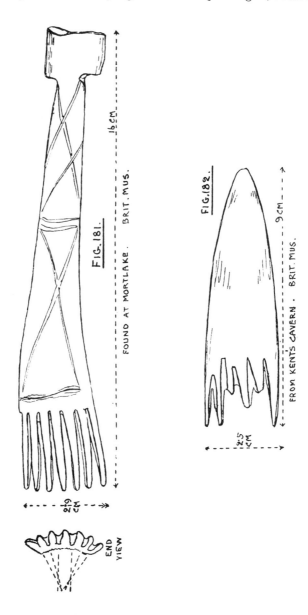

FIG. 181.

FOUND AT MORTLAKE . BRIT. MUS.

16 cm

END VIEW

FIG. 182.

FROM KENTS CAVERN . BRIT. MUS.

9 cm

specimen the teeth are apparently slightly more varied in shape, but owing to three of them having been broken off it is not possible to say anything as to their original length. They are both very rough on the concave side due to the exposure of the spongy interior portion of the bone of which they are made. The ornamentation is crude, consisting of crossed lines, etc., and the common circled dot.

E. T. Stevens, in referring to the collection of this class of tool from the Highfield Pit Dwelling, Salisbury, now in the Blackmore Museum, Salisbury, speaks of them as " bone and horn (red deer's antler) combs," and says regarding them: " These implements closely resemble some in recent use by the Eskimos for scraping fat, etc., from the backs of skins. The Esquimaux tools are made of wood, with the sharp claws of birds lashed to them. In the Christy Museum there are examples of these ; in the same collection there is a Basuto tool used for a similar purpose, the short thick teeth of which are of iron, bound to a wooden handle with twisted fibres. These modern implements help us to understand the use of the ancient tools."[1] From this one must infer that Stevens thought these instruments might have been made for skin-dressing purposes, although he was too cautious to commit himself.

Eleven years later Pitt-Rivers, in describing the excavations at Mount Caburn Camp, near Lewes,[2] devoted several pages to a description and record of finds of these tools in various parts of England, referred to Stevens' comparison between them and the Eskimo and Basuto skin-scraping tools[3] and said of one of them : " the seven teeth in this comb are blunt and rounded at the points, showing that it could not have been employed for combing the hair, and may possibly have been used for driving the weft against the cloth in weaving ; the association of such combs in the broch [Pictish tower] of Burrian, where fifteen of them were found, with seven rubbing bones or calendering implements made of the jawbones of whale, and used for smoothing the web after it is woven, appears to confirm this opinion as to their use."[4] He spoke of another comb found in the island of Bjorko and continued : " It was believed to have been used in weaving ribbon, and was ornamented with the dot and circle pattern. The small looms in which ribbons are woven are still in use in Norway and parts of Sweden ; a drawing of one from Dr. Hazelius's museum of native utensils at Stockholm is annexed. (See cut). It is $1\frac{1}{2}$ feet in length, and 8 inches high ; the ribbon is about 2 inches wide, and the comb of wood that presses up the woof has numerous teeth. As the bone combs under consideration have seldom more than ten teeth, some other system must have been employed than that in vogue in Norway. They may also have been employed in combing flax or wool."[5] In the cut he gives an illustration of a modern Norwegian ribbon loom, which in all probability, has long since out-distanced any loom that may have been in existence when the toothed instrument we are discussing was in use, so that the tentative

---

[1] *Flint Chips*, London, 1870, pp. 64-65.
[2] *Archæologia*, xlvi, 1881.
[3] *Ib.*, p. 10.
[4] *Ib.*, p. 11.
[5] *Ib.*, p. 11.

comparison cannot hold good. He also gives illustrations " of four deer-horn combs of like form from Greenland, in the Ethnographical Museum at Copenhagen ; they have ten, eight, eight, and seven teeth respectively, and are said to be used for combing flax."[1] Unfortunately, Pitt-Rivers omits to note that flax does not grow wild, if at all, in Greenland, hence it is not likely that the natives required a tool for combing it. One gathers from his statements that he favoured the opinion that these instruments were beaters-in.

We now come to the Glastonbury Lake Village explorers, Messrs. Bulleid and Gray, who found a large quantity of these implements at this settlement. After stating that, as recently as 1872, opinions were divided as to the purpose of the tools, they continue : " But it is now generally accepted that they were employed by the weaver in the upright loom for pushing home the weft (or woof) worked in by a shuttle, and so closing up the threads of the woven fabric—an operation absolutely essential in all kinds of looms. This process is now carried out in the horizontal loom by the swinging sley. These early weaving combs, therefore, served the same purpose as the reed, lay, or batten of our own time."[2] Here we have a positive opinion as to the function of this peculiar tool, of which many illustrations are supplied. To support their view the authors give us a diagrammatic representation, showing how the teeth of the tool, fitting into the warp dents, act both as a warp spacer and a beater-in. On examining their illustrations of these tools, one is struck at once by the difference in the number of teeth—they vary from five to fourteen—and by the wide diversity in the form of the dents ; most are naturally wedge shaped, but with varying depths on one and the same tool, a variation which also applies to the dent head which, in a few cases, runs to an extremely acute angle and in others is somewhat more open.

In the Mortlake tool in the British Museum (Fig. 181), owing to the rounded surface of the bone having been left in its natural state, the teeth are not in the same plane, being built on a base concavo-convex in section, hence only the centre portion of the tool beats-in when the convex side is used and only the outer teeth beat-in when the concave surface is used. Then, also owing to the rounded nature of the bone, the sides of the dents converge towards a point about an inch or so on the concave side, instead of every one being parallel to its neighbour, so that, when used to beat-in the warp threads are drawn out of position. As a matter of fact, on trying to use this tool (a facsimile in so far as possible of the Mortlake specimen) instead of obtaining the flawless result illustrated in Bulleid and Gray's diagram (Fig. 183), I got the distorted result shown in my illustration (Fig. 184). But not only was the warp alignment distorted, but in beating-in considerable friction was evoked between the

[1] *Ib.,* p. 12.

[2] A. Bulleid and Harold St. George Gray, *The Glastonbury Lake Village.* Glastonbury Antiquarian Soc., 1911, 1, pp. 268-9.

warp and the teeth. The curved base of the teeth of the beater-in brings the outer dents closer together and their sectional lines instead of remaining parallel become radii, converging at a point on the concave side, thence we have not only the negligible slight occasional contact between every warp and the teeth on either side of it, but a very close contact indeed. In fact so great is this that it amounts to a positive

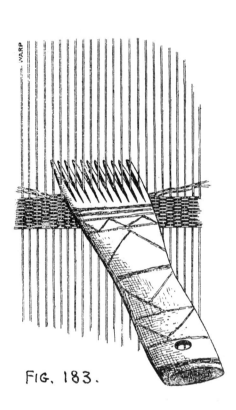

FIG. 183.

FROM BULLEID & GRAY'S GLASTONBURY LAKE VILLAGE

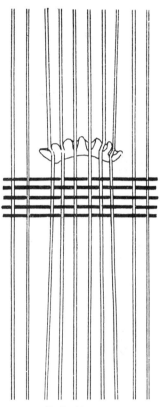

FIG. 184.

WARP DRAWN OUT OF POSITION BY THE DENTS (OF THE MORTLAKE QUASI WEAVERS' COMB) WHICH PARTLY CONVERGE AT POINTS ON THE CONCAVE SIDE INSTEAD OF RUNNING PARALLEL TO ONE AN--OTHER.

hindrance to the work, necessitating greater hand pressure and decided wear and weakening of the warp. With a beater-in on which the teeth are in a straight base, even if not well spaced, the friction is minute, but, of course, the greater the number of teeth the greater the friction, and this is again intensified with a concavo-convex base. I obtained the same results on warp placed horizontally or vertically, and I may add I tried the original tool on some primitive looms in the British Museum,

133

which trial first raised my doubts as to the alleged use of the implement. Anyone can make these trials for himself.

In any case Bulleid and Gray's diagram is an anachronism, for if the Glastonbury people used warp-weights (and I think the perforated articles the authors call loom weights are such, and not net sinkers) then the vertical loom with these weights was in use. As is well known in these upright warp-weighted looms, the weaving proceeded from above downwards, hence the beating-in must be from below upwards. In Bulleid and Gray's diagram the beating-in is from above downwards—what the Glastonbury people, it is safe to say in the present state of our knowledge, never did on warp-weighted looms. The Copenhagen Museum's Scandinavian warp-weighted loom, as illustrated by Montelius, and the Iceland loom illustrated by Olafsson,[1] show a sword or dagger-shaped beater-in and do does the Icelandic loom in the Reykjavik Museum.[2] In the manufacture of the Chilkat blanket on an *apparently* warp-weighted loom, the author mentions no such a tool as a beater-in, saying the whole of the work is done by the fingers.[3] In their diagram, too, the authors make the tool flat, forgetting their statement that in cross section these tools " are for the most part concavo-convex."[4] By this oversight they overcome the difficulty inherent where the dents converge to one point on the concave side instead of being in parallel lines. The very acute angle at which some of the dents terminate must cause the yarn to get wedged and on the withdrawal of the tool some of the warp will get lifted up and so displace the work, thereby encompassing the very object which is most to be avoided. As a minor objection the roughness of the concave portion of the bone where the cancellous tissue of the horn or bone has not been removed is liable to catch both warp and weft and disarrange them.

The chief objections to the use of the " combs " as beaters-in of the weft are :—

1. The concavo-convex base of the teeth, which—

    (*a*) Cause the warp to be displaced laterally and thereby
    (*b*) Cause excessive friction.

2. The great irregularity in the width of the dents culminating in the acuteness of the dent heads which have the tendency to " bite " the warp and obstruct working.

---

[1] Both reproduced in *Ancient Egyptian and Greek Looms*, pp. 34 and 35.

[2] Daniel Bruun, *Faeroerne, Island og Gronland paa Verdensudstillingen i Paris*, 1900, Kjobenhavn, 1901, p. 25.

[3] Emmons, *op. cit.* pp. 343-4.

[4] Of the numerous illustrations of these articles with which they supply us only two, numbered B232 and H33, Pl. xlvi, appear usable as weft beaters-in. *Op. cit.*, p. 270.

Hence the conclusion one comes to is that the tool is unsuitable for beating-in the pick and was, therefore, not intended for that sort of work.   There may be a few of these instruments which can be made to do the work, but in that case it will be because the obstacles I point out are by chance minimised or absent.

The so-called Egyptian weaver's comb, with its parallel semi-teeth, is quite a different article from the Glastonbury tool, and as I have practically shown,[1] is of

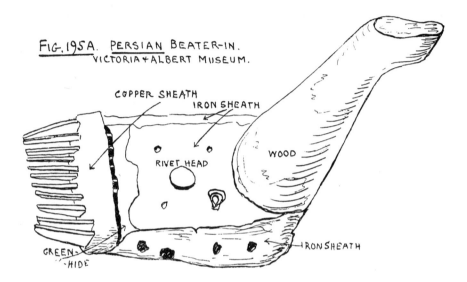

FIG. 195A. PERSIAN BEATER-IN.
VICTORIA + ALBERT MUSEUM.

COPPER SHEATH
IRON SHEATH
RIVET HEAD
WOOD
GREEN-HIDE
IRON SHEATH

no use for weaving purposes on a warp-weighted loom.   A similar article to the Egyptian tool, but with full teeth like the Roman " comb " found at Fort Donald, which may have been used on a loom, is the Wilton carpet-weavers' beater-in.   As Wilton carpet weaving is an introduced trade, this tool was no doubt introduced with it and can have no connection with the Glastonbury article.   We have toothed beaters-in in India, Persia, Asia Minor, North Africa, etc., but some are almost perfectly straight like the Wilton tool, others are bent like the Aures tool (Fig. 91B) or Lahore tool (Fig. 194A), and others again are doubly bent as the Persian (?) tool (Fig. 195A). As to the alleged comb carved on a panel of a bench-end in Spaxton Church, Somerset,[2] all the tools there represented seem to me to be cloth-finishing and not cloth-weaving implements, and the article specially referred to by Bulleid and Gray has the appearance of the brush used for putting on paste on certain cloths.   But in any case a woodcut illustration of a church wood carving is hardly sufficiently accurate evidence on which to base or support a theory.

---

[1] " Bishop Blaize, Saint, Martyr, and Woolcombers' Patron," *Proc. Soc. Antiq. Lond.*, 1914, and *Bankfield Museum Notes*, 2nd Ser., No. 6, Fig. 11, p. 31.

[2] J. R. Green, *A Short Hist. of the English People*, illustr. Ed., Lond., 1904, p. 783.

According to the discoveries made on the sites of the Swiss Lake Dwellings,[1] anything like the Glastonbury tool seems to have been very rare—it is possibly mentioned twice. On the site of the Stone Age village of Moossee, where no record is made of metal articles, nor spindle-whorls, nor warp-weights, although it is highly probable weaving was carried on there, " a comb of yew-wood, 2½ inches (or 7·6 cm.) broad and nearly 5 inches (or 12·7 cm.) long " was found.[2] It is depicted as flat, with nine very regular teeth or eight dents, which, if intended for beating-in, would indicate about 3·2 warp to the inch,[3] so that the author appears to be correct in stating that it " was probably used as a comb for keeping up the hair." From the Nussdorf site of a somewhat later age, where no metals were met with, but plenty of spindle-whorls and warp-weights, " three combs were also found, made out of a flat piece of stag's horn."[4] The teeth of the one specimen depicted look decidedly like those of the Glastonbury instrument and the tool is shown to be convex on one side at least. It appears to be about 3 inches (or 7·6 cm.) long and about 1¼ inches (or 3·2 cm.) broad, with seven very irregular teeth or six dents, which, if intended for beating-in, would indicate about 4·8 warp to the inch, but, as in the Glastonbury specimens, the dent-heads run out to such a fine point that great difficulty must have been experienced with them if they were used as beaters-in. No such articles are recorded to have been found at Robenhausen (also a Stone Age site of nearly the same age as that of Nussdorf, with traces of bronze and copper), where, no doubt owing to special circumstances, a large amount of evidence as to the existence of weaving has been found in the form of charred cloth. At this place was recovered an article described as a wooden knife about 6 inches (or 15 cm.) long, which has all the appearance of a sword beater-in,[5] as we see it in Peru, etc. The evidence of the Swiss Lake Dwellings is thus not very illuminative for this our enquiry. There is, however, a big field still open for any investigator who wishes to take up the study of the Swiss Lake Dwellers from the weaver's point of view.[6]

There are two tools which bear a close resemblance to the Glastonbury so-called weavers' comb, viz., the Pueblo Indians' toothed beater-in and the Eskimo skin softener (Fig. 186). The Glastonbury and Pueblo instruments are much alike superficially and hence they have been easily confounded. I have in Bankfield Museum two specimens of the American Indian toothed beater-in, one

---

[1] Ferd. Keller, *The Lake Dwellings of Switzerland*, transl. by J. E. Lee, 2nd Ed., Lond., 1878.

[2] *Op. cit.*, p. 38, Pl. v., No. 21.

[3] I cannot find in the illustrations of cloths given by Keller any in which two warp are laid through one dent as in the Sangier cloth, Fig. 132A, although naturally this does not mean that the Swiss Lake Dwellers did not use this method occasionally.

[4] *Op. cit.*, p. 119, Pl. xxviii, No. 8.

[5] *Op. cit.*, p. 52, Pl. x, No. 2.

[6] On Pl. xli, No. 9, of Keller's work there is an illustration of an article which is described as a " shuttle," but there seems to be no reason why it should not be called a whistle.

from the Navajos, well finished with seven teeth, obtained by exchange from the American Museum of Natural History, and the other (Fig. 185) a rather crude production with six teeth, given me by Miss Mary A. Owen, obtained from some Apaches, who no doubt adopted it from the Pueblo peoples. The Pitt-Rivers (Oxford) Museum's specimen dated 1884, obtained from the Zuni by Prof. Moseley, is a rough specimen provided with eight teeth. Miss B. Freire-Marreco, of Somerville College, has in her possession a sketch of a fairly well finished one with five teeth seen in use by a Hopi Indian at Oraibo.

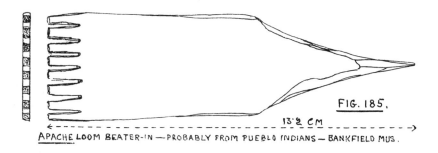

FIG. 185.

13·2 CM

APACHE LOOM BEATER-IN — PROBABLY FROM PUEBLO INDIANS — BANKFIELD MUS.

All these tools are more or less flat and have a more open dent head than the Glastonbury instruments. Washington Mathews gives two small illustrations of the toothed beater-in, which he calls a reed-fork ; both are depicted as being flat in section.[1] Miss B. Freire-Marreco, who has studied weaving amongst the Pueblo Indians, very kindly writes me in answer to my enquiries :

" As far as I know, toothed beaters-in used by the Hopi Indians and their Tewa neighbours are more or less flat in section, except for the teeth themselves, which are tapered in section as well as in plan. I know of no direct evidence for the tool being indigenous or introduced into the Pueblo area. On the one hand, the pre-Conquest sites of the Pueblo area have not, as far as I know, yielded specimens of this or any other weaving tool which can be possibly identified as such ; on the other hand it seems highly improbable that the Indians should have derived this beater-in from the Spaniards, who introduced the European hand-loom with swinging reed or batten. On the whole I am disposed to consider the toothed beater-in as indigenous to America. Its use appears to be associated characteristically with the vertical blanket loom of the Pueblo and Navajo Indians, which (in spite of Otis Mason's opinion) I believe to be an entirely native development, rather than with the belt loom (rigid heddle) which Otis Mason shows to be probably derived from European models, for, although the miniature beater-in in the Pitt-Rivers collection is associated with a belt loom, I have always seen the weft of the belt loom pushed home with the fingers without any tool, whereas with the vertical blanket loom the toothed beater-in seems to be indispensable."

[1] *Rep. Bureau of Ethnol.*, 1881-2, 1884, p. 382.

Miss Freire-Marreco's experience practically confirms what Washington Mathews tells us about the tool. Although he depicts two of these toothed beaters-in in connection with a belt loom he does not mention it when describing the act of weaving on such a loom, but he does mention its use when describing the act of weaving on a Navajo blanket loom.[1] He tells us the toothed beater-in and sword beater-in are used in ordinary procedure, but that the latter has to be discarded when the cloth is so far finished as not to allow of its insertion any further, for it is too broad for the space left, but into which the toothed beater-in, owing to its narrow flat section, can easily be pushed. It must be remembered, as already explained (p. 20), that the Navajos and other American weavers have a distinct method of beginning their wefting at both ends, or of weaving right up to the warp beam. The toothed beater-in is consequently an instrument specially designed to assist a certain method more or less indigenous to America, and hence it most probably is also indigenous and cannot be the same tool as the Glastonbury and other prehistoric so-called weavers' combs, quite apart from the fact that a concavo-convex implement would not answer the purpose for which the Navajo toothed beater-in is necessary.

I think the above shows clearly that the Navajo toothed beater-in and the Glastonbury alleged " weavers' comb " are quite distinct from each other, and that the latter was not used by weavers for beating in the weft. Such being the case, what was the function of the Glastonbury tool ?

The accompanying illustration (Fig. 186) represents some bone tools used by the Eskimo in skin dressing. They differ from the alleged weavers' combs found in Britain in one respect only, namely, in that a portion of the whole cylindrical bone is used instead of a portion of the longitudinal section ; in all other respects they agree, so that it seems fairly evident that the peculiar implements we are dealing with were used for skin-dressing and that Stevens, in making the suggestion referred to, was correct in his surmise. I think, in addition, that the opinion that these instruments were skin-dressing tools is supported by the fact that so many of the teeth are broken, which would not occur with ordinary beating-in of weft, but would, and does occur in the hard work the tool is put to in skin-dressing. The natives of South Africa formerly used very hard thorns wherewith to do the work, now they use iron spikes or nails.[2]

While we are told that the Glastonbury folk kept a considerable number of cattle and sheep and goats and, from the quantity of articles made of red deer antlers, we may infer they killed red deer, no mention is made in the Glastonbury Records of the dressing of skins, or of the use of skins in any way. The natives must have had skins, but no doubt all traces of any skin or leather have disappeared long ago and hence the explorers are unable to make any record of them. The natives may not

[1] Op. cit., p. 382.
[2] E. Vaughan-Kirby : " Zululand Skin-Dressing." Man, Mar., 1918, 23.

have used skins for clothing purposes, for there is plenty of evidence, in the existence of warp weights and spindle whorls, that they were weavers, but the skins being there must have been made use of and here we have tools which were adapted for dressing the skins and were no doubt used for that purpose. This, so to speak, absence of first hand evidence of the existence of dressed skins or leather in any form has also, I venture to think, misled Bulleid and Gray as regards the functions of certain pieces of worked wood which they say, are " presumably parts of looms of appliances for making textile fabric."[1] In Plate LV they show some of this wood

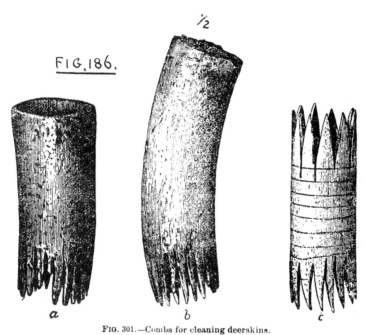

½

FIG. 186.

a b c

FIG. 301.—Combs for cleaning deerskins.

FROM JOHN MURDOCH'S ETHNOLOGICAL RESULTS OF THE POINT BARROW EXPEDITION .IXTH. ANN. REP. BUREAU OF ETHNOLOGY. 1892. P. 301.

made up into a frame as found *in situ*. I am quite unable to make it serve in any way as a *loom* frame. But if we complete it by merely filling in the twenty small round holes in the frame with pegs protruding on the upper surface we obtain what looks like a skin-dressing frame, such as we find in a primitive form among the Eskimo of Bering Strait as illustrated by Edward Wm. Nelson,[2] which is an advance on the Zulu method of ramming strong pegs into the ground as explained by E. Vaughan Kirby in his paper on Zulu skin-dressing.[3] The two lugs in the Glastonbury frame would not hinder the work of dressing the skin in any way, but

[1] *Op. cit.*, p. 340.
[2] xviii *Ann. Rep. Bur. Ethn.*, Part I, p. 116.
[3] *Man*, Mar., 1918, p. 36.

apart from them the frame is similar in almost every respect to the frame known as a *herse*, used by leather manufacturers in the middle of last century.[1]

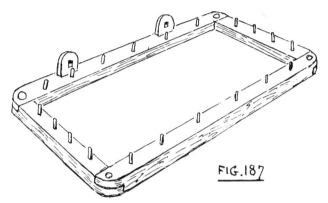

FIG. 187

A POSSIBLE SKIN STRETCHING FRAME RECONSTRUCTED FROM THE ILLUSTRATIONS OF SUPPOSED LOOM PARTS ON P.L. LV. OF BULLEID + GRAY'S GLASTONBURY LAKE VILLAGE.

[Although I am obliged to dissent from some of the conclusions arrived at by Messrs. Bulleid and Gray, I hope my so doing will not be construed into any want of appreciation of the excellent piece of work they have accomplished.]

12. CONCLUSION :—ORIGIN AND DISTRIBUTION.

It will be fitting to close these studies with some remarks on the Evolution and Distribution of the Looms which have been under discussion in so far as there is any evidence to go upon.

Tylor, in discussing the question as to how any particular piece of skill or knowledge had come into any particular place where it is found, says : " Three ways are open, independent invention, inheritance from ancestors in a distant region, transmission from one race to another ; but between these three ways the choice is commonly a difficult one."[2] It is a very difficult one. Not the least obstacle to coming to a decision is the apparent simplicity of the loom in its earliest stages, for so simple does it appear that one is tempted to pronounce judgment forthwith and say such a simple tool must have suggested itself to mankind in the remotest times and hence have had a common origin. On the other hand, being such a simple tool it must have been invented many times over. Origin or Invention must precede Distribution or copying and is consequently more remote and obscure than distribution, which in most cases is so obvious that it tends to increase the obscurity of origin.

---

[1] Chas. Tomlinson, *Illustrations of Useful Arts and Manufactures*, London [1858 ?], p.61, Fig. 267.

[2] *Researches into the Early Hist. of Mankind*, London, 1878, 3rd Ed., p. 376.

Origin or Invention requires predisposing circumstances and material, self-control and imagination or mental alertness—the slow progress being due to the fact that the alert minded portion of the community is generally in a minority. We do not know much about the circumstances conducive to an improvement, nor are we sure to understand the working of the primitive man's mind when brought into contact with circumstances favourable for innovation. It is also an open question whether among primitive peoples every invention is made " into some predetermined form," as maintained by Otis Mason.[1] It may not be possible to get but one result, and in so far the form must be predetermined. Otherwise it can hardly be correct to say the form is predetermined in the inventor's mind. Many inventions are haphazard results ; others are results quite different from what was anticipated. Some inventors have only a very hazy notion of what the result is likely to be, whatever object they may have in view, and others again are very clear as to the actual form the invention is to take. Mason is probably nearer the mark when he contradicts himself a few lines lower down, and states that every invention commences " with the relief of discomfort through a happy thought by means of some modification or new use of a natural object." How far physical necessity or advantage urged early man forward is difficult to estimate, for, apart from such pressure, there is the desire to outshine one's fellows—a feeling, perhaps, as strong among primitive peoples as amongst the more highly civilised.

The lower the state of development of a people the lower will be the inventive or progressive situation, so that while we get simple inventions in early times we get simple and complex ones in later times—the reason being that in the later times man has a store of fore-knowledge on which to premeditate. It is not likely to be the case often that man would have the opportunity to invent a complex tool a second time, for complex tools appear late, i.e., when transport, contact, etc., have been quickened, but he can go on inventing new applications of a principle. John Kay invented the Fly Shuttle in 1733, and in so doing adopted the same principle as is used by the Loyalty Islanders in their javelin propeller Ounep (Kennedy Collection, Bankfield Museum), of which he could not have known anything. At the start a principle will, generally speaking, not be clear to man, and he will experiment— often, unconsciously, thinking he is doing ordinary work—until out of a hazy conception the principle manifests itself to him. We have seen this in the development of the flying machine. Scheele and Priestley independently discovered oxygen, and Priestley did not know what he had discovered. Darwin and Wallace both formulated the theory of the Origin of Species independently. Professor E. H. Parker has shown that the Chinese script was evolved quite independently of any other.[2]

---

[1] *Origin of Inventions*, 1895, p. 15.

[2] " The Origin of Chinese Writing," *Journ. Manchester Egyptian and Oriental Society*, 1915-16, p. 61.

In this connection it may not be out of place to bring forward an analogous instance derived from the lower animals. Thus, we find in Insects that the faculty of producing silk has been independently acquired in certain cases. Among the caterpillars of the Lepidoptera, silk is the product of a pair of tubular glands which open into the mouth. The silk is liberated at the apex of an organ known as the spinneret. Among certain of the Neuropterous insects, on the other hand, the silk is derived from glands opening into the hind intestine, the threads being discharged through the anus. Whether the silk is identical from the chemical standpoint in all cases is very doubtful, but this point does not invalidate the analogy. The function of the silk is the same, both among the Lepidoptera and Neuroptera, viz., that of forming the cocoon in which the insect may transform into the pupal stage. When we find the principle of independent evolution among lower forms of life we may expect it among higher forms. Hence we have the two methods of shed making—that of " Carton " weaving (*Tissage aux Cartons, Brettchenweberei*) and that of heddle or ordinary weaving.

It is not necessary that inventions of a like nature should all be made at once. A Halifax man, named Hemingway, secured in 1909, a copyright for a design for an anti-splash sink, that is a sink on which the sides at the top are made to bend over inwards in order to prevent water splashing over. He told me he was led to this invention by noticing the mess made in his scullery by water being splashed on to the floor, and was much astonished when informed later on that the Ancient Egyptians made pots with a rim which had the same effect—probably the fore-runner of the vase—which I could show him in Bankfield Museum. Whether this rim was intended by the Egyptian potter to prevent oversplashing when in use we cannot say. It is possible, but, doubtful whether there may be found in nature two independently evolved organs of like forms which have quite different functions. However this may be, the Nicobar Islanders use a back scratcher, *Kanchuat-ok*, which may be correctly likened to a spindle and whorl, the whorl being made out of a disc of coco-nut shell—the specimen referred to is in the E. H. Man Collection in Bankfield Museum. The islanders are, or were, innocent of twisted or spun fibre, using finely split cane instead.

The loom is after all only the frame upon which a principle, weaving, is worked out, and judging from what has been observed above, there is considerable reason for the supposition that it may have been invented more than once.

When I was in Queensland some years ago, 1878-1884, I found it was common knowledge among bushmen that where the aborigines had been unable to procure European-made axes or knives they had turned to broken glass bottles and converted these into suitable cutting tools. Not only did they make use of old bottles, but on the overland telegraph routes in the early days they used to climb the poles to appropriate the insulators for use as cutting tools, thereby frequently interrupting communications. In some cases they produced from old glass bottles an implement far superior to anything they had ever possessed before. An illustration of such a glass tool is given by Balfour in *Man*, 1903, No. 35. On the other hand, when, in a Reserve, other aborigines were shown how to set potatoes, they dug them up at night and ate them. This apparently contradictory conduct may be explained thus : in the first instance the aborigines had been accustomed to make cutting tools out of certain minerals, and when they found a new suitable material they proceeded to make use of it for the same purpose. In the second instance they knew nothing about setting tubers, or had only the haziest notions as regards planting of seed for the purpose of collecting a crop later on[1]; the prospective benefit of the setting appeared too far fetched to their limited experience and want of self-control, and they vitiated any possible results of their labour by satisfying a more immediate want. The presence of the new material with a cognate pre-existing industry and some mental alertness enabled them to produce an improved article which was a step forward, an invention, while on the other hand a new material without a cognate pre-existing industry failed to excite their imagination or control. In other words, in the discovery of making glass tools they were assisted by a preceding step, while in the potato setting they had no such assistance. To us, with our vast and slowly acquired experience in the matter, the planting of foodstuffs is a reasonable and necessary proceeding, but to these aborigines it was a huge jump from gathering ripe fruits in certain localities at certain seasons, and they had not the power of mind or imagination to carry them so far or to realise what the new action involved. It is when sudden innovations are sprung upon a primitive people that they are staggered —their mental equilibrium gets upset because they are accustomed to go forward slowly step by step. This anti-innovation attitude cannot therefore be attributed to conservatism or obstinacy, as Professor G. Elliot Smith thinks.[2] He points out how dividual this attitude is with many peoples in various parts of the world, which incidentally makes it a fair example of the " similarity of the working of the human mind," with which opinion, however, he does not agree.[3] This attitude is the same as that to which Professor Flinders Petrie refers when summarising the results of his investigations on Egyptian Tools and Weapons and calls the " remarkable resisting power " of certain countries against the introduction of the commonest types. It proves how strong and independent were the civilisations affected.[4] This attitude,[5]

---

[1] Although on the W. Coast of Australia, according to information given me by the late well-known explorer, A. C. Gregory, the aborigines when digging up *ajuca* or *wirang* (wild yams) re-inserted the head so as to be sure of a future crop (See " Origin of Agriculture," by H. Ling Roth, *Journ. Anthrop. Inst.*, xvi, 1887, p. 131).

[2] " Ships as Evidence of the Migrations of Early Culture," *Journ. Manchester Egyptian and Oriental Society*, 1915-16, p. 81.

[3] *Ibid.*, p. 97.

[4] *Tools and Weapons*, London, 1917, p. 65.

[5] In very late or much more civilised times the attitude becomes an economic one. " A peasant does not adopt a new process easily, because he cannot afford risks, while experience

then, which while opposing contact retards distribution, must have considerable effect in permitting the internal slow growing-up of new forms : in other words, must be a stimulus to local origins.

In the case before us the stone tool-making industry paved the way for the glass tool industry. This was possibly only taking a first step, but every step, however small, is the forerunner of others, which when they have reached a certain stage are used as landmarks to indicate that a new position or a new form has been attained, which is designated the Origin or Invention of the article involved.

As mentioned at the outset of these papers, the consensus of opinion amongst those who have given attention to primitive weaving is that weaving is indebted for its origin to basketry and matmaking. I am more inclined to think that, owing to the difficulty of making the foundation or centre of baskets, not bags, basketry becomes a side-issue leaving mat-work in a more direct line of evolution from wattle-work. The evolution proceeded probably with inter-twined branches to form a breakwind, developing into fairly regular wattle, or more pliable material was brought into use, and then a finer and softer material was used by which mats were produced, the work in the meanwhile dividing into plaiting and plain up and down woven matwork, until for the latter a frame was laid out and the origin of the loom was attained. In the meanwhile spinning in the form of making twine had been discovered and the spun yarn ultimately ousted the non-spun filament used in the matmaking. But long before any such progress could be recorded there were the wattle- or mat-work industries which paved the way. These industries are wide-spread amongst primitive or unrisen people, and the instances are rare in which such people have not yet begun to utilise the natural facilities of their surroundings in order to produce this class of work. Where they have not done so they might have proceeded to do so later on had they been left alone, but the impediment to estimating such a possibility is our want of knowledge of the continuous life of such people, for as soon as we or other races come in contact with them the continuity of their life is broken, the slow step by step Invention ceases, and Distribution with difficulty takes its place.

Throughout the Solomon Islands there is an important matwork industry, not so much of value from the utilitarian point of view as from the decorative point of view, for these people are endowed with considerable artistic feeling. Ornament with them is almost an essential to their well-being. The same material which is used in their decorative matwork is used as warp and weft in their loom. This loom is one step forward from their method of making decorative tubular matwork. In making this one forward step they still continue to produce the same tubular matwork, but now fabricate it on a specially designed frame—in other words, they

shows that an old mode continues to pay." (H. Ling Roth, " Arbère : A Short Contribution to the Study of Peasant Proprietorship," *Journ. Statistical Soc. London*, March, 1885.)

have now invented the loom.   As already pointed out, the Solomon Islanders owe
nothing to the far-travelled Santa Cruz loom ; the whole arrangement, details and
method of working of the two frames are dissimilar, and all they have in common is
the qualification that they are both looms.   Although not so advanced as the Santa
Cruz article, the Buka (Solomon Islanders) loom is clearly an article in the course
of being built up as already explained, and the people who make it are, in spite of
their savagery, very alert-minded.   But the loom is only just a loom and still lacking
that essential of all further developments, the heddle, which naturally points to
recent evolution, which again precludes inheritance from ancestors in distant regions.

We have, then, the predisposing or preparatory industry in the form of decorative
matwork, carrying with it the existence of suitable material, the mental alertness
of the people, the extremely primitive form of the loom, freedom from exotic
influence, and clean progressive workmanship, which all tend to point to a local
independent origin of the Buka loom.

The case of the African vertical mat loom is somewhat different.   We do not
know how long this loom may have been in existence.   The Bushongo have a
tradition that a certain chief of one of their allied tribes taught his tribe how to
weave, and the other tribes learnt the art from this one.   Commenting on this,
Torday and Joyce[1] consider that the art was learned before the people settled where
they are now to be found.   Assuming a possible migration from Ancient Egypt,
or assuming a more immediate contact of the Ancient Egyptians and the Bantu-
speaking peoples dating back some 4,000 years or so, we should expect variations to
suit the genius of the adopting party as well as to suit local conditions, and we
should expect also to find that the greater the difference between the two, or any two,
civilisations, the greater will probably be the variations at the end of the long lapse
of time and migration or break of contact.   Between Penelope's loom, as illustrated
on the Chiusi skyphos and the Scandinavian looms in the Copenhagen or Reykjavik
Museums—with a period of remote ancestry amounting to about 2,600 years—
there is a greater difference than between the Pacific type of loom as it exists on both
sides of that ocean, although there is a closer connection between the Ancient
Greeks and the Scandinavians than there is between the Ancient Mexicans and
those Indonesians who use the Pacific form of loom.

The points in common between the Ancient Egyptian and African mat loom
are verticality and the possession of heddles, and, in so far as the working result is
concerned, the absence of selvedge in the earlier Egyptian productions.   The
Egyptian weaver used balls of weft hanging above his head from which he drew his
lengths of filament as required, much as the Eastern rugmaker does at the present
day : he used no spool in so far as is yet known.   The African weaver makes use of
an early specimen of the needle form of weft carrier.   The Egyptian used fine spun

[1] *Op. cit.*, p. 183.

linen yarn ; the African uses non-spun split palm leaf filament. The African heddles are but two steps removed in development from the first use of fingers in the raising of the warp, and neither in width nor in length can the African loom-woven mat compare with that of the Ancient Egyptian cloth. These Africans have succeeded in producing artistic patterns as well as pile cloth,[1] results to which the Egyptians never seem to have attained, the whole being, of course, based on non-spun filaments. Some of the looms show improvements in detail over others, that is, they show various stages in building up.

If the African loom is the outcome of remote contact with the Ancient Egyptian, one must ask how is it that both Egyptian forms have not been preserved, for the African to-day only uses the vertical and semi-vertical (or semi-horizontal) form and not the horizontal form ? Also, are the divergences and persistences what we should expect to find ? As shown above, what we have reason to expect does not occur. Instead of searching so far afield, let us see what wide local influences may have accomplished. There exists among the Bushongo and the closely connected tribes an intensive and extensive mat-making industry, which owes its existence, continued if not original, to the natural abundance of the material provided by the Raphia palm leaf. Specimens of this mat-work when brought to Europe by the Torday Expedition showed it to be of very considerable merit, and as such proved a surprise to African students, who could not fail to see that here was a hitherto unknown African people which had attained to a comparatively high state of civilisation. The work is also, necessarily, in every respect quite a contrast to the degenerate products obtained from the West Coast.

We have the matwork industry which, with the concomitant suitable material, could pave the way for further developments, the still early form of the loom, the remoteness from a possible prototype coupled with the wide divergences exhibited between the two looms and the clean progressive workmanship, all of which tend in the direction of an independent local building-up rather than to a possible remote exotic ancestry.

The Egyptian wall paintings of the eleventh dynasty, of at least 2000 B.C., illustrate the horizontal form of loom.[2] Those of the eighteenth dynasty illustrate the vertical form of loom. In the interval between the earlier and later representations there was the Hyksos invasion as well as the Syrian campaigns of Thotmes III, with the result that alien people in large numbers began to make their appearance in the country. The Hyksos introduced horsemanship[3] and long

---

[1] In the Manchester Museum there is a specimen of pile cloth with an old label attached indicating it to be ancient Egyptian ; but Miss W. M. Crompton informs me that the cloth is probably Coptic, and not earlier than A.D. 300.

[2] *Anc. Egyptian and Greek Looms*, p. 41. On line 4 from top, *for* horizontal loom *read* vertical loom.

[3] Breasted, *A History of the Ancient Egyptians*, 1908, p. 184.

range archery. It is possible that these aliens may have introduced the vertical loom. Or the second form may have developed out of the first, for we have indeed an intermediate form of loom among the above-mentioned far-off people, the Bushongo, which rests at an angle of 45° on the ground, the weaver squatting under the incline. There is no evidence to go upon beyond the fact that after a considerable turmoil in Egypt we find a vertical loom where previously only a horizontal loom was depicted. However this may be, Egypt gives us evidence of the existence of looms which goes back to extremely remote times, and the evidence is not outdistanced by that of the Sumerian tablets with their records of weaving work given out. The Egyptians were a progressive people : they had a big mat-making industry and *inter alia* at one period possessed bedsteads of which the foundation was strong twisted filament interlaced at right angles on a rectangular frame.[1] There is, however, a considerable gap between their matwork with its usual non-spun filament, and the linen cloths which have come down to us with their fine-spun filament, and so far we are unable to fill up the gap, but as this is in the line of evolution it presents no great obstacle. If the difference between the two looms is as great as that between the Buka loom and the Santa Cruz loom, both as regards form and development, then we can safely say, perhaps, that the Ancient Egyptians invented a loom, which fact would coincide with Professor Petrie's view of the want of distribution between the two peoples. We have, however, no clue whatever as to the form of the Sumerian loom.

There is a broad, flat, semi-toothed, handled instrument,[2] generally spoken of as a weaver's comb, mentioned on p. 135, which appears to have made its first appearance in Egypt in Roman times, for it is not discoverable in any of the numerous Ancient Egyptian weaving scenes. As we see it, it is, of course, not in its original form, and I believe some writers, including myself, have imagined it to be the forerunner of the reed. It could not have been in use with warp-weighted looms. It may have come into use with the introduction of the cloth beam. I now think this so-called comb, this beater-in, was a special device evolved with the invention of pile rugs or carpets where the old sword beater-in would have the tendency to undo the " knotting." On the other hand, I have outlined above the whole course of the evolution of the reed from a notched stick to the complete article, an evolution which can be seen in full operation in Indonesia at the present day. The reed originated in an effort to keep the warp regularly spaced, and the effort ended up not only in thoroughly accomplishing the desideratum, but, outstripping the inceptive idea, made a perfect beater-in as well. There may

[1] See the Specimen of a bedstead of the early part of the First Dynasty in the Manchester Museum.

[2] *Ancient Egyptian and Greek Looms*, Fig. 22.

possibly have been an embryo reed in the surmised Egyptian warp spacer,[1] but, as mentioned when dealing with it, we are quite without proof from India or Indonesia which would enable us to say it has travelled from Egypt. The cloth made on these looms is very broad and long, and something more than laze-rods is wanted to keep the warp threads spaced, and hence the invention or perhaps a migration from Egypt in later times. In Nigeria we find a peculiar warp spacer (Fig. 98), used with the vertical cotton looms, which may be an embryo reed.

The Pueblo Indians appear to have invented a special toothed instrument for pressing in the warp, originating in the necessity to overcome the difficulty created by their method of beginning to weave at both ends of the warp, which again may be due to their not using heading-rods.

The shuttle traces its origin to a transverse winding of the weft yarn, which tends to make spool and weft together thicker in diameter than when the yarn is wound round the spool longitudinally. At first sight one would think such a clumsy contrivance a poor sort of invention, for it hindered rather than helped the pickmaking. Its very clumsiness, however, led to the adoption of an easing sheath, which paved the way for the evolution of the modern shuttle. This evolution can be seen in various stages in Indonesia at the present day. Ancient Egypt has so far only produced balls of yarn, and at that stage, to the best of our knowledge, the Egyptians left it when their country was overrun by the Romans.[2]

The rectangular loom frame appears to have sprung from the bringing together and combination of two separately evolved parts of looms, viz., a frame for supporting reed and heddles and their harness to a frame supporting a warp beam. This was in Indonesia. It may very possibly have grown up in another way farther west, which perhaps accounts for its wide distribution in Asia Minor and the shores of the Mediterranean, etc. Its isolated presence on the West Coast of Africa I have explained as due direct to European influence.

The Ainu have invented a special form of warp spacer, and the Chinese, Japanese, and Koreans make use of a C-spring arrangement for raising the heddles, a form of harness which is peculiar to themselves.

From the above it is clear enough that we have a fair amount of evidence to the effect that some looms and various portions of others have been more or less

---

[1] *Ancient Egyptian and Greek Looms*, Fig. 23.

[2] In his very useful book, *Tools and Weapons*, Lond., 1917, Professor Flinders Petrie illustrates on Pl. lxvi, Fig. 127, a weft carrier which he calls a Roman shuttle. As the illustration is too small for examination, he has very kindly sent me particulars from which I gather that the article is an eighteenth-century English shuttle with exotic decoration. Professor Petrie has since further informed me that he does not know the provenance of this shuttle, which was purchased by him. The other weft carrier which he illustrates, Fig. 126, which he calls a shuttle, is a spool, and not a shuttle. Speaking presumably of Egyptian and Roman weft carriers he says, on p. 53, " Shuttles are rather rare." Unfortunately, so far, none at all have been found.

invented *in situ*, and do not owe their existence to distribution or copying or from contact with other people, nor from remote ancestry. Of other looms, without our being able to indicate their origin, we can safely say that where they are now met with they have found their way by migration or contact. Such looms are the African Fixed Heddle Loom, the African Pit Treadle Loom, and the African Horizontal Narrow Band Loom, all probably of Asiatic origin. As regards this Narrow Band Loom it has gone through so many changes during its migration that, compared with its prototype, it is almost unrecognisable. The warp-weighted loom was in evidence in Ancient Greece and also in the Swiss Lake Dwellings and England at the commencement of the Bronze Age. We have records of it in Scandinavian Saga in the eleventh century, and it was probably in use amongst the northern peoples several hundred years before then. It has lasted in Iceland until quite recent years, and may possibly still be worked there by the natives of the sparsely inhabited northern coasts, according to information I received, before the War, from Shetland fishermen who had been there.

To sum up, it seems almost as certain as can be ascertained from such limited studies as these that some looms are of independent invention, others are an inheritance from ancestors in a distant region, and others again have been transmitted from one race to another.

ADDENDUM.

Students having asked me to explain the wefting of the looms, Figs. 80 and 81, I give here the method by means of which I have been able to weave on the principle they typify.

I.—THE MADAGASCAR LOOM.—A pick is made in the shed as shown in No. 1. The shed stick A is moved up to the fixed heddle, as shown in No. 2, and a pick made. A is moved back to its position as in No. 1 and the original shed is re-formed.

II.—THE A-FIPA LOOM.—The position of the shed stick B, in No. 1, is obtained by placing it as shown in No. 3, where this shed stick carries on the countershed to the fabric. When position No. 1 is obtained a pick is made in the countershed and B is withdrawn when the shed is formed as in No. 2 ; here another pick is made. Then position No. 1 is re-obtained by moving A up to the fixed heddle and carrying the countershed past the heddle by re-inserting B.

The A-Fipa weaver makes countershed and shed and then two picks, then countershed and shed and two picks again, and so on, while

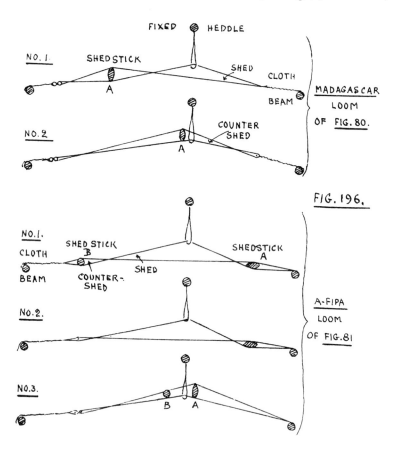

FIG. 196.

the Madagascar weaver follows the usual sequence of shed and pick, countershed and pick, shed and pick again, and so on. There are thus two methods of weaving on a fixed heddle loom.

INDEX

INDEX